Electricity
DeMYSTiFieD®

DeMYSTiFieD® Series

Electricity
DeMYSTiFieD®

Stan Gibilisco

Second Edition

New York Chicago San Francisco Lisbon London Madrid Mexico City
Milan New Delhi San Juan Seoul Singapore Sydney Toronto

McGraw-Hill books are available at special quantity discounts to use as premiums and sales promotions, or for use in corporate training programs. To contact a representative, please e-mail us at bulksales@mcgraw-hill.com.

Electricity DeMYSTiFieD®, Second Edition

1 2 3 4 5 6 7 8 9 0   DOC/DOC   1 7 6 5 4 3 2 1

ISBN 978-0-07-177534-2
MHID 0-07-177534-X

Sponsoring Editor	**Project Managers**	**Proofreaders**
Judy Bass	Joanna Pomeranz and	Don Pomeranz and
	Nancy Dimitry,	Don Dimitry,
Editing Supervisor	D&P Editorial Services	D&P Editorial Services
Stephen M. Smith		
	Composition	**Art Director, Cover**
Production Supervisor	D&P Editorial Services	Jeff Weeks
Richard C. Ruzycka		
	Copy Editor	**Cover Illustration**
Acquisitions Coordinator	Nancy Dimitry,	Lance Lekander
Bridget L. Thoreson	D&P Editorial Services	

To Tony, Samuel, and Tim

About the Author

Stan Gibilisco, an electronics engineer and mathematician, has authored multiple titles for the McGraw-Hill *Demystified* and *Know-It-All* series, along with numerous other technical books and dozens of magazine articles. His work has been published in several languages.

Contents

Introduction

This book can help you learn the principles of basic electricity without taking a formal course. It can also serve as a supplemental text in a classroom, tutored, or home-schooling environment. None of the mathematics goes beyond the high-school level. If you need a refresher, you can select from several *Demystified* books dedicated to mathematics topics. If you want to build yourself a substantial mathematics foundation before you start this course, I recommend that you read and study *Algebra Know-It-All* and *Pre-Calculus Know-It-All*.

How to Use This Book

As you take this course, you'll encounter plenty of multiple-choice questions written in standardized test format. You'll find an "open-book" quiz at the end of every chapter. You may (and should) refer to the chapter text when taking these quizzes. Write down your answers, and then give your list of answers to a friend. Have your friend tell you your score, but not which questions you missed. The correct answer choices are listed in the back of the book. Stay with a chapter until you get most of the quiz answers correct.

Three major parts constitute this course. Each part ends with a multiple-choice test. Take these tests when you've completed the respective parts and have taken all the chapter quizzes. The part tests are somewhat easier than the chapter-ending quizzes. The course concludes with a final exam. Take it after you've finished all the parts, all the part tests, and all the chapter quizzes. You'll find the correct answer choices for the part tests and the final exam listed in the back of the book.

With the part tests and the final exam, as with the quizzes, have a friend divulge your score without letting you know which questions you missed. That way, you won't subconsciously memorize the answers. You might want to take each test, and the final exam, two or three times. When you get a score that makes you happy, you can (and should) check to see where your strengths and weaknesses lie.

I've posted explanations for the chapter-quiz answers (but not for the part-test or final-exam answers) on the Internet. As we all know, Internet particulars change; but if you conduct a phrase search on "Stan Gibilisco," you'll get my Web site as one of the first hits. You'll find a link to the explanations on that site. As of this writing, it's **www.sciencewriter.net**.

Strive to complete one chapter of this book every 10 days or two weeks. Don't rush, but don't go too slowly either. Proceed at a steady pace and keep it up. That way, you'll complete the course in a few months. (As much as we all wish otherwise, nothing can substitute for "good study habits.") When you're done with the course, you can use this book as a permanent reference.

If you'd like to work with "hands-on" circuits that demonstrate some of the principles presented in this course, you can check out my book *Electricity Experiments You Can Do at Home*. If your interest extends beyond the subject matter here, you can take the more comprehensive course *Teach Yourself Electricity and Electronics*.

I welcome your ideas and suggestions for future editions of this work.

Stan Gibilisco

Electricity
DeMYSTiFieD®

Part I

Direct Current

chapter 1

A Circuit Sampler

Learning to read circuit diagrams is like learning to drive a car. You can read books about driving or diagram-reading, but you need practice before you feel comfortable. This chapter will give you a little diagram-reading practice. As you proceed through the course, you'll get a lot more!

CHAPTER OBJECTIVES

In this chapter, you will

- Interpret block diagrams.
- Recognize a few of the basic schematic symbols.
- See how devices interconnect to form a complete electrical system.
- Read some component-level schematic diagrams.
- Distinguish between series and parallel component connections.

Block Diagrams

When we look at a *block diagram*, we see major components or systems shown as rectangles, and we see interconnecting wires and cables drawn as straight lines. Some specialized components have unique symbols identical to the ones used in more detailed circuit diagrams.

Wires, Cables, and Components

The block diagram of Fig. 1-1 shows an electric generator connected to a motor, a computer, a hi-fi stereo system, and a television (TV) receiver. We portray each major component as a rectangle or "block." The interconnecting wires in this system comprise multiple-conductor electrical cords or cables. In this simple diagram, all of the cords and cables appear as single, straight lines that run vertically or horizontally on the page.

TIP *In the interest of neatness, we should always try to draw lines representing wires, cords, or cables either straight across the page or straight up and down. We should use diagonal lines only when a diagram gets so crowded that we can't show a particular section of a wire or cable as a vertical or horizontal line without "cluttering things up."*

Adding More Items

In Fig. 1-1, none of the lines cross each other. Imagine that we want to add a *voltmeter* to the circuit. (A voltmeter measures electrical *voltage*, which we can

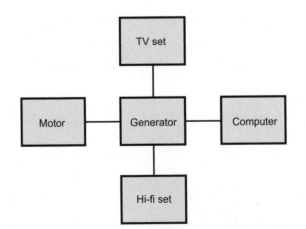

FIGURE 1-1 · A block diagram showing an electric generator connected to four common appliances.

imagine as a "force" or "pressure" that can cause electricity to "flow.") We want to install the meter so that we can selectively connect it between an earth ground and the input of any one of the four devices. This new component will make our diagram more complicated, so we'll have to let some of the lines cross.

Figure 1-2 shows how we can illustrate the addition of a voltmeter (symbolized as a circle with an arrow and labeled V), along with a four-way switch (the symbol for which appears just below the meter). The voltmeter has two terminals, one connected to the earth ground (the symbol with the three horizontal lines of different lengths) and the other connected to the central pole of the switch.

A Limitation

A block diagram can't portray all the details about a circuit. For example, we don't know whether the hi-fi set in Fig. 1-1 or 1-2 is simple or sophisticated. We don't know what features the computer has. We don't know whether the TV set connects to a cable network, a satellite system, or a simple wire antenna! Internal device details don't show up in block diagrams, such as Fig. 1-1 or Fig. 1-2.

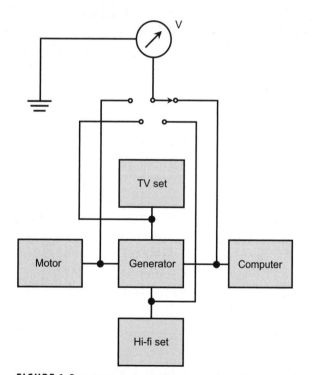

FIGURE 1-2 · Addition of a voltmeter, switch, and ground to an electrical system.

A block diagram doesn't necessarily tell us how many conductors a cable, when represented by a single line, actually has. Two lines emerge from the voltmeter in Fig. 1-2; each line represents a single-conductor wire. However, between the generator and the four major appliances, the interconnecting lines represent three-wire electrical cords, not single-conductor wires. The lines running from the switch to the inputs of each of the major appliances do represent single-conductor wires. Those wires connect only to the "electrically hot" conductor in each of the four three-conductor cords.

? Still Struggling

Why, you might ask, do we not show the three-wire cords as sets of three lines, all parallel to each other? The answer: We can, but block diagrams should, in general, show things in the simplest possible way. In a complete, detailed *schematic diagram* of the system of Fig. 1-2, we would need to show the three-conductor cords as sets of three lines running alongside each other. In a block diagram, we don't have to show that much detail. The lines in Figs. 1-1 and 1-2 show general electrical paths, not individual wires.

Connected or Not?

When we want to show that two wires or cables connect to each other where the lines come together in a diagram, we draw a black dot at the point of intersection. In Fig. 1-2, the dots represent connections between single wires running from the switch and the "hot" wires in the cords running to the motor, the computer, the hi-fi set, and the TV set.

What about points where lines representing wires from the switch *run across* lines representing cords between the appliances and the generator? The absence of a dot means that the wires or cables *do not* connect to each other. If we want to show that two intersecting lines represent electrically connected wires, we must draw a black dot where the lines cross. Figure 1-3A shows two lines crossing, but representing wires or cables that don't connect at the point of crossing. Figure 1-3B shows lines representing two wires or cables that do connect at the point of crossing.

Do you suspect a potential problem with the scheme shown at B? You should! What if the draftsperson puts a dot at the crossing point, but it's not

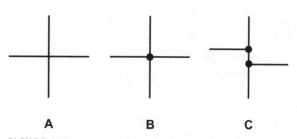

A   **B**   **C**

FIGURE 1-3 · At A, the two wires do not connect to each other. At B and C, they do connect. Drawing C shows the preferred way to portray connected wires when their lines cross in a diagram.

big enough for you to see easily? What if you fail to notice the dot even though it's big enough? Competent draftspeople overcome this trouble by using the scheme shown in Fig. 1-3C, the preferred way to portray two wires or cables that cross with a connection.

TIP   *When three or more lines come together at a point, it usually means that they're all meant to connect to each other, even if no dot appears at the point of convergence. Nevertheless, it's always good diagram-drawing practice to put a heavy black dot at any point where you want to tell your diagram-viewers that multiple wires or cables connect.*

PROBLEM 1-1

How else, besides the method at Fig. 1-3A, can we show that two crossing lines represent wires or that cables don't connect to each other?

SOLUTION

We can place a little "jump" or "jog" in one of the lines at the crossing point, as shown in Fig. 1-4. We'll sometimes see this portrayal in "ancient" texts and papers (those written before about 1970); we won't encounter it often in more recent documents. In this book, we'll use the method shown in Fig. 1-3A.

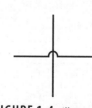

FIGURE 1-4 · Illustration for Problem 1-1.

PROBLEM 1-2

In the situation of Fig. 1-2, the voltmeter measures the electricity at the input to one (but only one) of the appliances. Which one?

✔ SOLUTION

Note the position of the arrowed line inside the switch symbol. It runs to the terminal that goes to the line representing the electrical cord for the computer. Therefore, the voltmeter in Fig. 1-2 connects to the computer's electrical input, and not to any of the other devices.

PROBLEM 1-3

Suppose that we want to show the direction in which the electricity "moves" in Fig. 1-2. How can we do that?

✔ SOLUTION

We can add large arrows pointing outward from the generator toward each of the major appliances, as well as toward and away from the voltmeter, as shown in Fig. 1-5. These arrows indicate that electricity "flows" from the

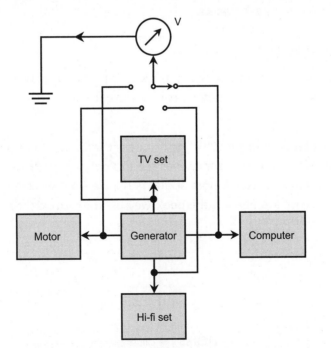

FIGURE 1-5 · Illustration for Problem 1-3.

generator to the appliances, and also from the generator through the volt-meter to ground. We must not confuse the large arrows with the small arrow inside the switch symbol. The small arrow tells us where the center contact of the switch connects, but it doesn't necessarily indicate the direction in which electricity "flows" through the switch.

Schematic Diagrams

Let's look at some of the symbols used in detailed diagrams of electrical devices as we examine some simple circuits. For a comprehensive table of symbols that professionals employ when they draw electrical and electronic circuit diagrams, refer to the appendix at the back of this book.

TIP   *You might want to start studying the appendix right now, and review it often. When you've completed this course, you can use the appendix as a permanent reference.*

Flashlight

A *flashlight* comprises a battery, a switch, and a light bulb. We connect the switch so that it can interrupt the "flow" of electricity through the bulb. Figure 1-6A shows a flashlight without a switch. Figure 1-6B shows the same flashlight with a switch added, allowing us to illuminate or extinguish the bulb at will.

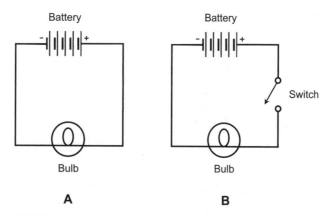

FIGURE 1-6 · At A, a battery and a bulb are connected together. At B, we add a switch to make a common flashlight. This drawing is also the subject of Problem 1-4.

Note that we connect the switch in line with (or, as engineers say, in *series* with) the bulb and the battery, rather than across (in *parallel* with) the bulb or the battery. In the series circuit, the electricity must pass through all three devices—the switch, the bulb, and the battery—if we expect the bulb to light up.

? Still Struggling

Schematic diagrams can vividly illustrate the difference between parallel and series connections among three or more components. For example, Fig. 1-6 shows a bulb, a switch, and a battery in series with each other. Fig. 1-7 shows the same three components connected in parallel with each other.

PROBLEM 1-4

In the circuit shown by Fig. 1-6B, should we expect the bulb to light up? If so, why? If not, why not?

✔ SOLUTION

We should not expect the bulb to glow. The fact that the arrowed line "misses" the terminal to the bulb tells us that switch is in the *open position* (or "off," so that it constitutes an *open circuit*). An open switch can't conduct any electricity. Because the switch, the battery, and the bulb appear in series with each other, a single break at any point in the circuit will prevent electricity from "flowing" at any other point.

PROBLEM 1-5

What will happen if we place the switch in parallel with the battery and bulb, as shown in Fig. 1-7, rather than in series with the battery and the bulb? What fundamental mistake would that arrangement represent?

✔ SOLUTION

In the scenario of Fig. 1-7, we can expect the bulb to glow because the battery connects directly to the bulb. If we close the switch, however, things

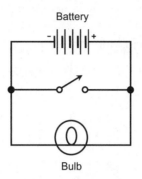

Battery

Bulb

FIGURE 1-7 • A bulb, a switch, and a battery connected in parallel with each other. This drawing is also the subject of Problem 1-5.

get complicated. In that situation, the battery and the bulb will both experience a direct *short circuit*, in which the end terminals directly connect to each other. The switch will "hog" all the electricity from the battery, leaving little or none for the bulb. The switch will conduct perfectly (or almost perfectly) while the bulb will have significant *resistance* (opposition) to the "flow" of electricity.

TIP   *The circuit of Fig. 1-7 shows an example of a circuit with a serious engineering flaw. We never want to let a perfect short circuit exist across a source of electricity such as a battery, even by accident.*

WARNING!   *A prolonged short circuit can cause chemicals to boil out of a battery, resulting in the leakage of hazardous materials into the environment. Some batteries can rupture or explode under such conditions. Even if a catastrophe like that doesn't occur, the circuit wires might heat up so much that they ignite surrounding materials and start a so-called* **electrical fire.**

Variable-Brightness Lantern

Imagine that we find an electric lantern bulb designed to work with six volts (6 V) of *direct current* (DC) electricity. It will light up even if we provide it with somewhat less than 6 V, but as you might expect, it will glow less brilliantly than it would if we gave it the full 6 V. Figure 1-8 shows a battery connected

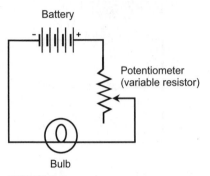

FIGURE 1-8 · A variable-brightness lantern. The potentiometer allows us to adjust the voltage that the bulb receives.

to a bulb through a variable resistor called a *potentiometer*. The zig-zags in the symbol tell us that the component is a resistor, and the arrow means that we can adjust the resistance. Let's suppose that the battery provides 6 V and the bulb will shine at its maximum brilliance when supplied with 6 V.

When we set the potentiometer for its lowest possible resistance (actually a direct connection), the bulb glows at full brilliance. When we set the potentiometer for its highest resistance, the bulb glows dimly or not at all, depending on how great that maximum resistance value happens to be. When we adjust the potentiometer for intermediate resistance values, the bulb shines more or less brightly. As the resistance goes down, the brightness increases; as the resistance goes up, the brightness decreases. If we choose a potentiometer having the correct amount of maximum resistance, we'll get a lantern whose brilliance we can adjust to any level we want.

PROBLEM 1-6

What will happen if we connect the potentiometer in parallel with the light bulb and battery, rather than in series with it? Why does that arrangement represent a bad idea?

SOLUTION

With the potentiometer at its maximum resistance, the bulb will shine at its brightest. As we reduce the resistance of the potentiometer, the bulb will get dimmer because the potentiometer will "rob" some of the electricity intended for the bulb. If we set the resistance too low, the potentiometer will burn out. If we set the resistance all the way down to zero, we'll short out the battery, creating a dangerous situation.

Multiple-Bulb Circuit

Suppose that we want to connect five light bulbs across a single battery and have each bulb receive the full amount of electricity from the battery. We can accomplish this task by connecting the bulbs and the battery in parallel with each other, as shown in Fig. 1-9. We'll encounter this sort of circuit in cars, trucks, travel trailers, and small boats. This arrangement contains no switches, so we can expect all the bulbs to glow constantly.

TIP *In Fig. 1-9, we don't label the symbols with component names. You should know what they represent by now. In standard operating practice, engineers rarely label individual schematic symbols with their functional identities. The symbols themselves convey that information!*

PROBLEM 1-7

If one of the light bulbs in the circuit of Fig. 1-9 burns out so that it conducts no electricity (as if it were an open switch), what will happen?

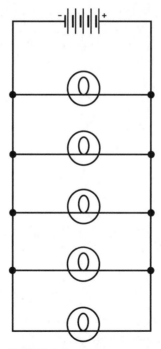

FIGURE 1-9 · A five-bulb circuit. The symbols lack labels because we already know what they represent. This drawing also relates to Problems 1-7 and 1-8.

☑ **SOLUTION**

The bad bulb will go dark. The other bulbs will all keep shining because they will all still receive the full amount of battery electricity.

PROBLEM 1-8

If one of the light bulbs in the circuit of Fig. 1-9 shorts out, what will happen?

☑ **SOLUTION**

The battery will experience a direct short circuit. All the bulbs will go dark because the short circuit will consume all the available battery electricity. Fortunately, light bulbs almost never short out when they fail. Instead, they open up, as described in Prob. 1-7.

PROBLEM 1-9

How can we add a switch to the circuit of Fig. 1-9, allowing us to turn all of the bulbs on or off simultaneously?

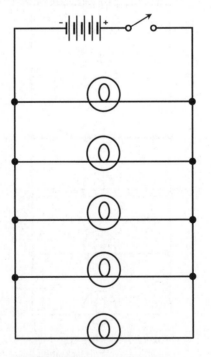

FIGURE 1-10 · Illustration for Problem 1-9.

✔ SOLUTION

Figure 1-10 shows how we can accomplish this task. We place the switch next to the battery. When the switch opens, it interrupts the electrical path to the entire set of bulbs.

TIP *In the circuit of Fig. 1-10, we place the switch next to the positive battery terminal. If we put the switch next to the negative terminal instead, the circuit will operate in exactly the same way.*

PROBLEM 1-10

How can we add switches to the circuit of Fig. 1-9 so that we can switch any single light bulb on or off, independently of all the others?

✔ SOLUTION

We insert five switches into the circuit, with one switch next to each bulb, as shown in Fig. 1-11. When we open a particular switch, it interrupts the electrical path to the adjacent bulb, but does not interrupt the path to any other bulbs.

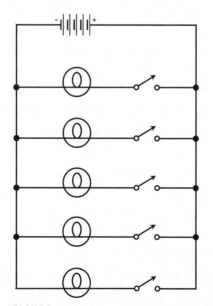

FIGURE 1-11 · Illustration for Problem 1-10.

More Diagrams

We can do plenty of things with a battery, a few light bulbs, several switches, and some potentiometers. The following paragraphs should help you get used to reading schematic diagrams of moderate complexity.

Universal Dimmer

We can add a potentiometer to the circuit in Fig. 1-11 so that we can adjust the brightness levels of all the bulbs simultaneously. In Fig. 1-12, the potentiometer acts as a universal light-brilliance control (often called a *dimmer*). The word "universal" means that the brilliance control affects all the bulbs. The electricity, which follows the wires (straight lines), must pass through the potentiometer to "flow" from the battery through any single bulb, and back to the battery again.

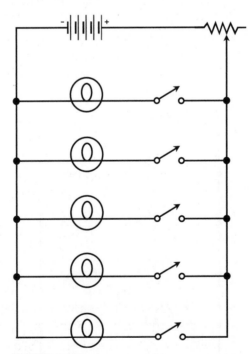

FIGURE 1-12 • A circuit in which we can individually switch five light bulbs and simultaneously adjust their brilliance.

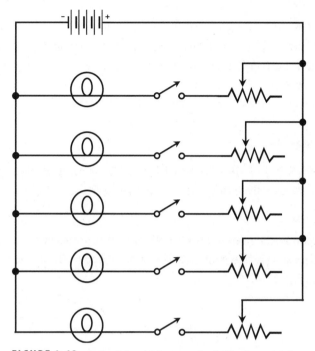

FIGURE 1-13 · A circuit in which we can individually switch five light bulbs and individually adjust their brilliance. This drawing also relates to Problem 1-11.

Individual Dimmers

Figure 1-13 shows a circuit similar to the one in Fig. 1-11, except that in this case, each bulb has its own individual potentiometer. Therefore, we can adjust the brightness of any single light bulb, as well as switching any single bulb on or off.

PROBLEM **1-11**

If we change the setting of one of the potentiometers—say, the second from the top—in the system of Fig. 1-13, the resistance change will affect the brilliance of the corresponding bulb. What about the brilliance of the other bulbs? Will the adjustment of the second potentiometer from the top affect the brightness of, say, the second bulb from the bottom?

SOLUTION

No. Each potentiometer in Fig. 1-13 affects the brilliance of its associated bulb, but not the brilliance of any other bulb. The dimmers in this circuit are all independent.

PROBLEM 1-12

Can we do anything to the circuit of Fig. 1-13, so that all the lights can be dimmed simultaneously, as well as independently?

 SOLUTION

Yes. We can add a universal-dimmer potentiometer as we did in Fig. 1-12, in addition to the five that already exist in Fig. 1-13, obtaining the circuit shown in Fig. 1-14.

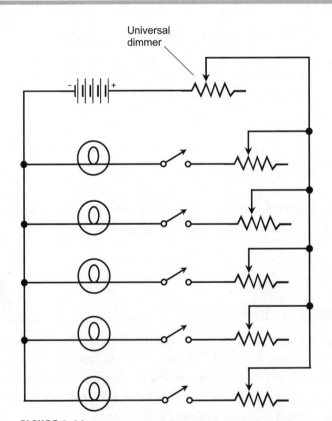

FIGURE 1-14 · Illustration for Problem 1-12. All the unlabeled potentiometers are independent brilliance-adjustment controls.

QUIZ

This is an "open book" quiz. You may refer to the text in this chapter. You'll find the correct answers listed in the back of the book.

1. Figure 1-15 shows two ways in which we can interconnect the ends of four wires called W, X, Y, and Z. Which of the following statements holds true in practical terms?

    A. In the scheme at A, all possible pairs of wires connect directly, but in the scheme at B, only the pair W-to-Y connects directly.
    B. In the scheme at A, all possible pairs of wires connect directly, but in the scheme at B, only the pairs X-to-Y and W-to-Z connect directly.
    C. In the scheme at A, all possible pairs of wires connect directly, but in the scheme at B, only the pairs W-to-X and Y-to-Z connect directly.
    D. The scheme at A shows the same situation as the scheme at B; either way, all possible pairs of wires connect directly.

2. Figure 1-16 shows five resistors called R_1 through R_5, all connected to a battery. The circuit also contains a voltmeter called V and an *ammeter* (a device that measures electrical *current* and tells us how "fast" the electricity "flows") called A. As shown here, the five resistors are connected in

    A. series with each other.
    B. parallel with each other.
    C. neither series nor parallel with each other.
    D. a combination of series and parallel with each other.

3. As shown by Fig. 1-16, meter A (which, in theory, conducts as well as a short circuit would) is connected specifically to tell us the electricity that "flows"

    A. through R_5, but not through R_1, R_2, R_3, or R_4.
    B. through the combination of R_1, R_2, R_3, and R_4, but not through R_5.
    C. through the combination of R_4 and R_5, but not through R_1, R_2, or R_3.
    D. out of the battery, but not through any of the resistors.

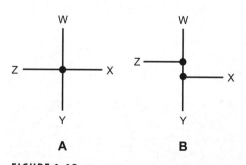

        A   B

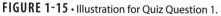

FIGURE 1-15 • Illustration for Quiz Question 1.

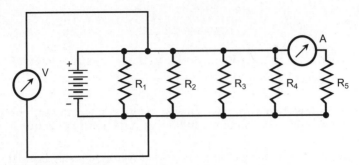

FIGURE 1-16 · Illustration for Quiz Questions 2 through 4.

4. **As shown by Fig. 1-16, meter V (which, in theory, conducts as poorly as an open circuit would) is connected specifically to tell us the electricity that exists**

    A. across R_1, but not across R_2, R_3, R_4, or R_5.
    B. across the combination of R_2, R_3, R_4, and R_5, but not across R_1.
    C. across any or all of the resistors R_1 through R_5.
    D. across the battery, but not across any of the resistors.

5. **Suppose that we connect six light bulbs in series, and then we connect the whole series combination to a battery. After an hour or so, two of the bulbs burn out simultaneously. What happens to the rest of the bulbs?**

    A. They stay lit, but they all get a little dimmer.
    B. They stay lit, but they all get a little brighter.
    C. They stay lit at the same brilliance as before.
    D. They all go dark.

6. **Suppose that we connect six light bulbs in parallel, and then we connect the whole parallel combination to a battery. After an hour or so, two of the bulbs burn out simultaneously. What happens to the rest of the bulbs?**

    A. They stay lit, but they all get a little dimmer.
    B. They stay lit, but they all get a little brighter.
    C. They stay lit at the same brilliance as before.
    D. They all go dark.

7. **In the circuit of Fig. 1-17, component X allows us to**

    A. switch the entire device on or off.
    B. control the brightness of the bulb.
    C. vary the resistance of the bulb.
    D. vary the voltage that the battery produces.

8. **In the circuit of Fig. 1-17, component Y allows us to**

    A. switch the entire device on or off.
    B. control the brightness of the bulb.
    C. vary the resistance of the bulb.
    D. vary the voltage that the battery produces.

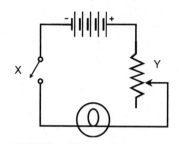

FIGURE 1-17 · Illustration for
Quiz Questions 7 and 8.

9. Look back at Fig. 1-2 on page 5. How many of the four devices (TV set, computer, hi-fi set, and motor) can we connect directly to the voltmeter *at the same time*?

  A. All four
  B. Three
  C. Two
  D. One

10. Suppose that we connect seven light bulbs in parallel with a large battery. All of the bulbs light up to full brilliance. If we remove five of the bulbs, what will happen to the amount of electricity that each of the other two bulbs receives?

  A. It will decrease.
  B. It will stay the same.
  C. It will increase.
  D. We need more information to answer this question.

chapter 2

Charge, Current, Voltage, and Resistance

Why can electricity do all sorts of things when we close a switch, and yet seem useless when we open that switch? What makes electricity manifest itself? What factors regulate its intensity? Let's investigate the nature of electricity by examining four of its major characteristics: charge, current, voltage, and resistance.

CHAPTER OBJECTIVES

In this chapter, you will

- Learn how electrostatic forces operate.
- Discover what causes electrical charge.
- Define and quantify electrical charge.
- Define and quantify electrical current.
- Define and quantify electrical potential, also known as voltage.
- Define and quantify electrical resistance.

Charge

For electricity to exist, we must have a source of *electric charge* that can manifest itself in two distinct and opposite ways. Scientists use the terms *positive* and *negative* (sometimes called *plus* and *minus*) to represent the two types of charge.

Repulsion and Attraction

The earliest electricity experimenters noticed that when they brought two objects near each other after "electrifying" them, those objects pulled toward each other (*attraction*) or pushed away from each other (*repulsion*). The so-called *electrostatic force* operates through the space between objects, just as two permanent magnets attract or repel depending on how and where we position them.

Electrostatic force and its "twin," *magnetic force*, can operate through air and other gases, various solids and liquids, and even a perfect vacuum (a fact that amazes some people). Two electrically charged objects attract if one has a *positive charge* and the other has a *negative charge* (Fig. 2-1A). If both objects have positive charges (Fig. 2-1B) or negative charges (Fig. 2-1C), they repel each other. The magnitude of the force, whether attractive or repulsive, depends on two factors:

1. The total, combined amount of charge on the objects
2. The distance between the centers of the objects

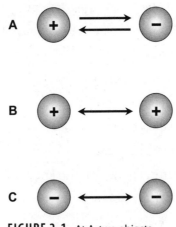

FIGURE 2-1 • At A, two objects attract each other if they have opposite charges. At B and C, two objects repel each other if they have like charges.

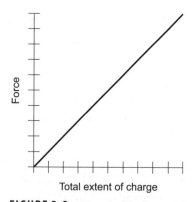

FIGURE 2-2 • When the total quantity of charge on two objects increases but nothing else changes, the force between them increases in direct proportion to the total charge.

As the total extent of the charge, known as the *charge quantity* (considered on the two objects taken together) increases, and if the distance between the centers of the objects doesn't change, the force between the objects increases in *direct proportion* to the total charge. Figure 2-2 shows a graph of this relation: force versus charge quantity.

As the separation between the centers of two charged objects increases, and if the total charge remains constant, the force decreases according to the *square* of the distance between their centers (Fig. 2-3). For example, if we double the

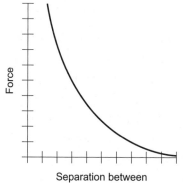

FIGURE 2-3 • When the distance between the centers of two charged objects increases but nothing else changes, the force between them decreases according to the square of the distance.

distance, the force goes down to 1/4 of its previous intensity. If we triple the distance, the force goes down to 1/9 of its previous intensity.

The Atom

For centuries, scientists have known that matter exists in tiny pieces or particles. We call those particles *atoms*. But that's far from the whole story. The more we study matter, the more complicated its structure seems to get. Physicists have discovered particles that behave like matter in some situations, and like energy in other scenarios. Sometimes things act like particles, and sometimes they act like waves. To understand the basics of electricity without getting into a lot of atomic physics, let's take a simplified view.

Individual atoms comprise smaller particles called *protons*, *neutrons*, and *electrons*. Protons and neutrons have incredibly small size (far smaller than any ordinary microscope can portray) and phenomenal density. We find them "clumped" together in a central mass called the *nucleus*. Electrons are much less dense than protons or neutrons, and they move around a lot more. Some electrons "orbit" a specific nucleus and stay there indefinitely, but they sometimes move from one nucleus to another. Protons and neutrons, in general, don't move from one atom to another (except in *nuclear reactions*, which won't concern us in this course).

Protons and electrons carry equal and opposite electric charges. Scientists consider protons as *electrically positive*, and electrons as *electrically negative*. These *charge polarity* definitions are arbitrary, having come about as a coincidental result of observations made long ago in simple experiments.

TIP *The charge quantity on any single proton, no matter where we find it, equals the charge quantity on any other single proton. Similarly, the amount of charge on any single electron equals the amount of charge on any other single electron. As far as we know, these two rules hold true throughout the universe. Neutrons never have any electrical charge. An atom carries a net positive charge if it has fewer electrons than protons, or a net negative charge if it has more electrons than protons. An atom carries no electric charge if, and only if, it has an equal number of protons and electrons.*

Electrons

The negatively charged subatomic particles—the electrons—sometimes remain "loyal" to a single atomic nucleus, but often they "roam free." For that reason, electrons are of particular interest in the study of electricity. An excess or deficiency of electrons on an object gives that object a *static electric charge*, also called an *electrostatic charge*. If an object contains more electrons than protons, then that object has a *net negative charge*. If an object contains fewer electrons than protons, then that object has a *net positive charge*.

In charged objects of reasonable size, we find a great many electrons. When you shuffle across a carpeted room on a dry day, you acquire an electrostatic charge that consists of millions upon millions of electrons that either accumulate on your body or else get drawn out of your body. When you imagine this situation, you might wonder how all those electrons can build up on (or drain from) your body without putting your life in danger.

Under some conditions, a buildup of electrostatic charge is harmless. But under other circumstances, such as when you stand in an open field during a thunderstorm, a truly gigantic charge—vastly greater than the charge you get from shuffling around on a carpet—can develop. If you acquire a massive enough charge, and then the charge difference between your body and something else (such as a cloud) suddenly equalizes, you can get electrocuted, burned to death, or both!

Units of Charge

We can consider the charge quantity on a single electron as the equivalent of one "unit." The charge quantity is the same for every electron we observe under ordinary circumstances, always has negative polarity, and constitutes what physicists call an *elementary charge unit* (ECU). In practical terms, that's an extremely small charge quantity. Engineers usually employ a unit called the *coulomb* to quantify electrical charge. In equations and other mathematical statements, we symbolize "coulomb" or "coulombs" as an uppercase letter C. One coulomb (1 C) equals approximately 6,240,000,000,000,000,000 ECU. A physicist or engineer would write this huge number using *scientific notation*, also called *power-of-10 notation*, to get the equation

$$1 \text{ C} = 6.24 \times 10^{18} \text{ ECU}$$

Based on this formula, we can write other equations such as

$$10 \text{ C} = 10 \times 6.24 \times 10^{18} \text{ ECU}$$
$$= 6.24 \times 10^{19} \text{ ECU}$$

or

$$0.001 \text{ C} = 0.001 \times 6.24 \times 10^{18} \text{ ECU}$$
$$= 6.24 \times 10^{15} \text{ ECU}$$

TIP   *In this book, we won't use scientific notation very often. If you're serious about studying electricity or any other scientific discipline, however, you ought to get comfortable with this method of rendering huge or tiny quantities. For now, in regards to electrical charge, let's remember that 1 C represents a moderate amount of electrical charge, often encountered in the real world.*

PROBLEM 2-1

Imagine two charged spherical objects. Assume that the charge is distributed uniformly throughout either sphere; it's not "clumped" at any particular point on the outside or inside of either object. Suppose that the left-hand sphere contains 1 C of positive charge, and the right-hand sphere contains 1 C of negative charge, as shown in Fig. 2-4A. This situation results in an electrostatic force of attraction, which we can call F, between the two spheres. Now suppose that we double the charge quantity on both the left-hand sphere and the right-hand sphere while not changing the distance between them. What happens to the force?

SOLUTION

The total charge quantity, represented by the product of the charges on the two objects, increases by a factor of 2 (doubling each charge amount) times 2 (because we have two spheres). Therefore, the force quadruples to $4F$, and continues to manifest itself as an attraction between the spheres, as shown in Fig. 2-4B.

PROBLEM 2-2

Imagine the same two charged spherical objects as in the previous problem. Assume, again, that the charge is distributed uniformly throughout either sphere.

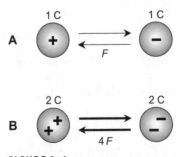

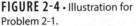

FIGURE 2-4 • Illustration for Problem 2-1.

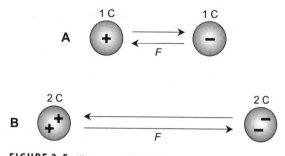

FIGURE 2-5 · Illustration for Problem 2-2.

The left-hand sphere contains 1 C of positive charge, and the right-hand sphere contains 1 C of negative charge (Fig. 2-5A), producing an attractive electrostatic force *F* between the two spheres. Suppose that we double the charge quantity on both spheres, and we also double the distance between their centers. What happens to the force?

✔ SOLUTION

The force remains the same. Doubling the charge on both spheres, but not changing the distance, increases the force by a factor of 4, as we've seen. But increasing the distance between the centers of the spheres causes the force to diminish by a factor equal to the square of the increase, which equals a factor of 2^2, or 4. In this particular case, the increase in force caused by the additional charge gets precisely "canceled out" by the decrease in force caused by the greater physical separation.

Current

When charged particles move through a physical medium, across a gap, or across a barrier, we get an *electric current*. Usually the current-carrying particles, known as *charge carriers*, are electrons. However, any moving charged object, such as a proton, an atomic nucleus, or an "electrified" dust grain, gives rise to an electric current.

Conductors

In some materials, electrons can move easily from one atomic nucleus to another. In other materials, the electrons can move, but only with some difficulty. In still other substances, we'll find it almost impossible to get electrons to move among atomic nuclei. Scientists and engineers define an *electrical*

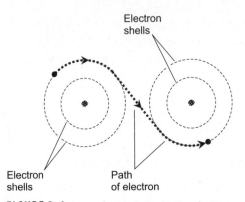

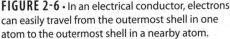

FIGURE 2-6 · In an electrical conductor, electrons can easily travel from the outermost shell in one atom to the outermost shell in a nearby atom.

conductor as a substance in which the electrons can move easily, so we don't have any trouble producing an electric current.

Pure elemental silver constitutes the best known electrical conductor among common materials at room temperature. Copper and aluminum are also excellent electrical conductors. Iron, steel, and most other metals constitute fair to good conductors of electricity. Some liquids are good conductors; pure elemental mercury (sometimes called "quicksilver") offers a representative (although toxic) example. Salt water is a fair conductor, but pure distilled water does not conduct very well. Gases, in general, make poor electrical conductors because the atoms or molecules are too far apart to allow a free exchange of electrons.

Figure 2-6 shows a situation where an electron (solid black dot) starts in orbit around a particular atom and ends up in orbit around another nearby atom. The electrons always orbit atomic nuclei at a specific average distance that takes the form of a sphere called a *shell*. (In Fig. 2-6, the thin dashed circles represent cross sections of electron shells.) When we observe a significant electrical current, the electron-movement process of Fig. 2-6 occurs among countless atoms, giving the appearance of a continuous current "stream."

? Still Struggling

The individual electrons in a conductor move somewhat like, but not exactly like, the molecules of water through a garden hose. If we're willing to accept a certain amount of oversimplification, we can represent a good electrical conductor as a large-diameter hose, and a poor conductor as a small-diameter hose. The amount of current corresponds to the volume of water (such as a cubic centimeter) flowing through the hose past any particular point per unit of time (such as one second).

Insulators

An *electrical insulator* is a substance in which electrons will not pass from atom to atom unless we exert a great effort to move them. Under ordinary conditions, insulators prevent current from flowing. Insulators, therefore, constitute poor conductors.

Most gases, being poor conductors, make good insulators. Glass, dry wood, paper, and plastics are other examples of excellent electrical insulators. Pure water also constitutes a decent insulator, although it conducts some current when it contains dissolved minerals (such as we often find in tap water or well water). Certain metallic oxides can act as good insulators, even if the metal in pure form makes a good conductor. Aluminum oxide is an example.

Engineers sometimes call a sample of insulating material a *dielectric*. This term, which translates roughly to the expression "two-electricities," arises from the fact that a piece of insulating material can keep a pair of electrically charged objects or regions with opposite polarity apart, preventing the flow of electrons that would occur if we let the objects come into contact or the regions overlap.

TIP   *When we keep two electrically charged objects or regions having opposite polarity separated with a dielectric object or medium, we call the pair of objects or regions an* **electric dipole.**

The Ampere

Scientists and engineers quantify current in terms of the number of charge carriers that pass a single point in one second (1 s). Usually, a great many charge carriers go past any given point in 1 s, even if the current remains small. For this reason, we'll almost never hear or read about current expressed directly in *ECU per second* (ECU/s), but we'll almost always hear or read about current expressed in *coulombs per second* (C/s). By definition, an electric current of 1 C/s represents an *ampere* (symbolized A), the standard unit of electric current.

Conventional versus Electron Current

Physicists consider the direction of an electrical current as going from a relatively positive charge pole to a relatively negative charge pole. This paradigm is known as *theoretical current* or *conventional current*. If you connect a light bulb to a battery, for example, the theoretical current flows out of the positive terminal and into the negative terminal. But the electrons, which carry the actual ECUs, flow in the opposite direction, from negative to positive. That's the *electron current*.

**PROBLEM 2-3**

Suppose that 6.24×10^{13} electrons flow past a certain point in 1 s of time. What's the current in amperes?

SOLUTION

Note that 6.24×10^{13} equals exactly 0.00001 C, so 0.00001 C/s flow past the point every second. We, therefore, observe a current of 0.00001 A.

Voltage

Current can flow only if charge carriers are "pushed" or "motivated" to move. The "push" can result from a buildup of electrostatic charges, or from a steady charge difference between two objects. When we have a positive charge pole (relatively fewer electrons) in one place and a negative charge pole (relatively more electrons) in another place not too far away, an *electromotive force* (EMF) exists between the two charge poles. We express this force, also known as *voltage* or *electrical potential*, in *volts* (symbolized V). We can also say that a *potential difference* exists between the two charge poles. Once in awhile, lay people might refer to voltage as "electrical pressure," but that's a technically imprecise expression.

Common Voltages

Household electricity in the United States, available at standard electrical outlets, has an effective voltage of between 110 V and 130 V; usually it's about 117 V (Fig. 2-7A). It's not a steady voltage, like the voltage from a battery; instead, it changes polarity several dozen times a second. A typical automotive battery has an EMF of approximately 12 V (Fig. 2-7B), and it's a constant voltage that never changes polarity. The charge that you acquire when walking on

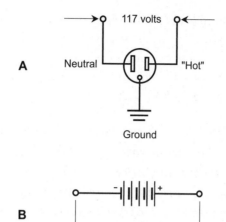

FIGURE 2-7 · At A, voltage at a common household utility outlet. At B, voltage at the terminals of an automotive battery.

a carpet with hard-soled shoes can sometimes rise to several thousand volts. Before a discharge of lightning, millions of volts build up between clouds, or between a cloud and the earth's surface.

Static Electricity

An EMF can exist between two charge poles without any flow of current. This situation occurs in most rain clouds, even if lightning never takes place. As you walk around on a carpet, a significant potential difference arises between your body and conducting objects connected to *earth ground* (such as copper cold-water pipes). An electrical potential exists between the terminals of an electrical outlet, even if we don't "plug in" anything. A voltage exists between the terminals of a lantern battery, whether or not we connect those terminals to a light bulb. When a voltage exists without an attendant flow of current, we say that we have *static electricity*. "Static" in this context means "not moving." Current can flow only through a conductive path between two points having different voltages.

An electrical potential difference, even a gigantic EMF, doesn't necessarily drive much current through a conducting medium. Consider your body after you've shuffled around on a carpet. A potential difference of several thousand volts might exist between your body and a massive, grounded iron heat radiator. This voltage seems deadly when you think about it in terms of sheer EMF. After all, it's far greater than the 117 V that appears at your wall outlets, and 117 V can electrocute you! However, not many coulombs of charge accumulate when you run around on a carpeted floor. Therefore, only a tiny amount of current flows through your finger when you touch a grounded metal object. You don't get a deadly shock, although the current surge might startle you.

TIP *Have you heard that current, not voltage, kills people? This statement holds true in a literal sense, but it oversimplifies the real situation. In theory, high voltage all by itself can't harm anybody. However, deadly current can flow only when sufficient voltage exists to "drive it." The current is directly responsible for electrocution, but a dangerous current can't flow without enough voltage to propel the charge carriers.*

? Still Struggling

When plenty of coulombs exist, you don't need much voltage to produce a lethal flow of current if you provide an excellent path for the flow of the charge carriers. This fact explains why no one should ever attempt to repair an electrical device when it's "plugged in." A "hot" utility outlet can pump charge carriers through the human body fast and long enough to cause injury or death.

WARNING! *Even a moderate EMF presents a genuine danger. When you're working around anything that carries significant voltage (more than about 12 V), have the same respect for it as you would have for a high cliff as you walk near the edge.*

Voltage versus Current

The current through an electrical component varies in direct proportion to the voltage supplied to the component (Fig. 2-8), as long as the characteristics of the component don't change. If we double the voltage, the current doubles. If the voltage falls to 1/100 of its original value, so does the current.

The straight-line relationship of Fig. 2-8 holds true only as long as a component always conducts to the same extent. In some components, the electrical *conductance* changes as the current varies. An electric light bulb is a good example. The conductance is different when the filament carries lots of

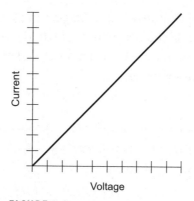

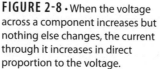

FIGURE 2-8 • When the voltage across a component increases but nothing else changes, the current through it increases in direct proportion to the voltage.

current and glows white hot, as compared to when it carries only little current and hardly glows at all.

TIP   *If you have my book* Electricity Experiments You Can Do at Home *(McGraw-Hill, 2010), you can find instructions for a test that demonstrates how the conductance of a lantern bulb varies depending on how much voltage you apply. Refer to the conclusion of Experiment DC-17 (pages 100 and 101).*

TIP   *Some electrical components and devices are designed to have constant conductance even when the voltages across them (and thus the currents through them) vary over a wide range. The most common examples are the* fixed resistors *that you find in electronic devices, such as radios and hi-fi amplifiers.*

PROBLEM 2-4

Why is it dangerous to operate electrical appliances with bare feet on a wet pavement or on bare ground, but much less dangerous if you wear rubber boots on your feet and dry leather gloves on your hands—even though the voltage is the same in both situations?

✔ SOLUTION

When your hands are exposed and you aren't wearing electrically insulating, waterproof shoes, a household utility voltage source can drive a lethal current through your body from an appliance to the ground or a wet floor. If you wear hole-free, dry leather gloves along with hole-free, dry rubber boots, you're safer because the voltage can't drive significant current through the gloves and boots! But no matter how good your gloves and boots might be, you should still make sure that you power down any electrical device before you service it.

WARNING! *Before you work on any electrical appliance or system, unplug the appliance from the utility outlet and shut off the outlet's supply of electricity at the fuse or breaker box.*

Resistance

Nothing conducts electricity perfectly. Even the best conductors have a little *resistance*, which we can define as opposition to electric current. Silver, copper, aluminum, and most other metals have low resistance. Some materials, such as carbon and silicon, have moderate resistance. Electrical insulators exhibit high resistance. Resistance is the opposite of conductance; they vary in *inverse proportion* with respect to each other.

The Ohm

Engineers express and measure electrical resistance in *ohms*. In some texts, you'll find the word "ohm" or "ohms" symbolized as the uppercase Greek letter omega (Ω). In this course, let's write the word out in full. As a component's resistance in ohms increases, current has more trouble flowing, given a constant voltage. Conversely, as we reduce the resistance in ohms with a constant voltage, we get more current.

In a conventional electrical system, we'll usually want to keep the resistance or *ohmic value* as low as possible. Resistance converts electrical energy into thermal energy (heat). This phenomenon is called *resistance loss* or *ohmic loss*. Usually, resistance loss represents energy gone to waste. Nevertheless, in some situations, we'll want to introduce a fixed or controlled resistance into an electrical circuit with the intent of regulating the current, or with the intent of deliberately generating heat.

When 1 V of EMF exists across a component with 1 ohm of resistance, we observe a current of 1 A through the component, assuming that the voltage source can deliver that much current. If we double the resistance while leaving the voltage unchanged, the current drops in half. If the resistance drops to half its original value, the current doubles. If the resistance increases by a factor of 5,

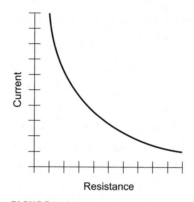

FIGURE 2-9 · When the voltage across a component remains constant, the current through it varies inversely with its resistance.

then the current decreases to 1/5 its previous value. If the resistance is cut to 1/5 its previous value, then the current increases by a factor of 5. As long as we keep the voltage constant, the current is proportional to the *reciprocal* of the resistance. In other words, the current varies *inversely* with the resistance (Fig. 2-9).

When current flows through a resistive component, the current gives rise to a potential difference across the component. As we increase the current through the component, assuming that the resistance remains constant, we get more voltage across that component. In general, the voltage varies in direct proportion to the current through the resistive object (Fig. 2-10).

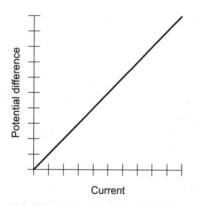

FIGURE 2-10 · When the current through a component increases but the resistance does not change, the potential difference across the component increases in direct proportion to the current.

Practical Resistance

Under normal circumstances, we'll never find a substance that completely lacks electrical resistance. However, circumstances need not always be normal! We can consider some materials as perfect conductors "for all intents and purposes" when we cool them down to temperatures near *absolute zero*, representing the coldest possible temperature in the universe. Then we observe *superconductivity*.

Just as there's no such thing as a perfectly resistance-free substance in ordinary everyday life, we'll never encounter anything that has truly infinite resistance, either. Even objects made of dry wood, plastic, glass, or air conduct electricity to some extent, although the effect is usually so small that we can ignore it. In some applications, engineers select materials on the basis of "how nearly infinite" a resistance value it has.

? Still Struggling

Engineers design certain components so that their internal resistances vary in specific and predictable ways depending on external conditions. A *transistor*, for example, might exhibit high resistance some of the time, and low resistance at other times. In such a component, the resistance can fluctuate between two different values thousands of times per second, facilitating the operation of high-speed switches and amplifiers like those found in personal computers and home entertainment equipment.

PROBLEM 2-5

Imagine a "black box" (let's call it component X) connected directly to a battery. Suppose that component X has a certain resistance. The voltage from the battery causes some current to flow through X. What will happen to the current if we double the battery voltage, and at the same time cut the resistance of component X in half?

✔SOLUTION

If the resistance of component X didn't change, then doubling the voltage would double the current. If the voltage across component X didn't change, then cutting the resistance in half would double the current. Because both things happen at once—the voltage doubles and the resistance drops to half—the current gets multiplied by a factor of 2 times 2, or 4.

QUIZ

This is an "open book" quiz. You may refer to the text in this chapter. You'll find the correct answers listed in the back of the book.

1. **When two objects have different electrical charges, we know** *for certain* **that**
    A. a current flows between them.
    B. a conductive path exists between them.
    C. an essentially infinite resistance exists between them.
    D. a voltage exists between them.

2. **Which of the following quantities varies in** *direct proportion* **to the resistance of a component, given a constant current through that component?**
    A. The voltage across the component
    B. The charge quantity in the component
    C. The charge polarity across the component
    D. The conductance of the component

3. **Which of the following quantities varies in** *inverse proportion* **to the resistance of a component, given a constant current through that component?**
    A. The voltage across the component
    B. The charge quantity in the component
    C. The charge polarity across the component
    D. The conductance of the component

4. **The flow of electrons in a current-carrying, straight wire runs**
    A. in the direction opposite the conventional current.
    B. from the more positive charge pole to the more negative charge pole.
    C. in the same direction as the conventional current.
    D. in circles at right angles to the conventional current.

5. **If we increase the resistance of a component by a factor of 4 while leaving the voltage across it unchanged, the current through that component**
    A. goes down by a factor of 16.
    B. goes down by a factor of 4.
    C. goes down by a factor of 2.
    D. remains the same.

6. **A charge quantity of 1.56 × 10¹⁸ ECU represents**
    A. 4.00 C.
    B. 2.00 C.
    C. 0.50 C.
    D. 0.250 C.

7. If 3.12×10^{18} ECU of charge carriers flow past a specific point in 1 s of time, the current through that point equals

    A. 0.250 A.

    B. 0.500 A.

    C. 0.667 A.

    D. 0.750 A.

8. If you shuffle around on a carpeted floor for a few minutes on a dry day while wearing hard-soles shoes, your body will likely acquire

    A. a deadly voltage.

    B. a deadly current.

    C. an electrostatic charge.

    D. All of the above

9. Which of the following particles always carries a positive charge?

    A. Atom

    B. Electron

    C. Proton

    D. Neutron

10. Imagine two electrons in a vacuum, close to each other but with nothing else nearby. If we increase the distance between the two electrons by a factor of 10, and if no other objects in the vicinity exert any effect on the electrons, the electrostatic force between them

    A. becomes 1/1000 as great.

    B. becomes 1/100 as great.

    C. becomes 1/10 times as great.

    D. does not change.

chapter 3

Ohm's Law, Power, and Energy

In this chapter, we'll look more closely at the ways current, voltage, and resistance relate to each other in simple DC circuits. We'll calculate electrical power and energy, based on current, voltage, and resistance. But first, let's learn how scientists express large multiples and small fractions of physical units.

CHAPTER OBJECTIVES

In this chapter, you will

- Use prefix multipliers to represent small or large quantities.
- Calculate voltage, current, and resistance values.
- Determine power levels based on voltage, current, and resistance values.
- Determine energy levels based on voltage, current, resistance, and elapsed-time parameters.
- Clarify the distinction between power and energy.

Prefix Multipliers

Physicists and engineers attach *prefix multipliers* to physical units to express large multiples or small fractions of those units. Table 3-1 lists the prefix multipliers commonly used in electricity and electronics.

For Voltage

You'll commonly encounter the *millivolt* (mV), which represents a thousandth of a volt. That is,

$$1 \text{ mV} = 0.001 \text{ V}$$
$$= 10^{-3} \text{ V}$$

The *microvolt* (μV) equals a millionth of a volt. Therefore

$$1 \text{ μV} = 0.000001 \text{ V}$$
$$= 10^{-6} \text{ V}$$

You'll occasionally see tiny potential differences expressed in *nanovolts* (nV), where

$$1 \text{ nV} = 0.000000001 \text{ V}$$
$$= 10^{-9} \text{ V}$$

Units larger than the volt exist. A *kilovolt* (kV) represents a thousand volts; that is,

$$1 \text{ kV} = 1000 \text{ V}$$
$$= 10^{3} \text{ V}$$

TABLE 3-1   Prefix multipliers used in the physical sciences. Each designator represents a specific power of 10.

Designator	Symbol	Multiplication Factor
pico–	p	0.000000000001 or 10^{-12}
nano–	n	0.000000001 or 10^{-9}
micro–	μ or mm	0.000001 or 10^{-6}
milli–	m	0.001 or 10^{-3}
kilo–	K or k	1000 or 10^{3}
mega–	M	1,000,000 or 10^{6}
giga–	G	1,000,000,000 or 10^{9}
tera–	T	1,000,000,000,000 or 10^{12}

One *megavolt* (MV) equals a million volts, so

$$1 \text{ MV} = 1,000,000 \text{ V}$$
$$= 10^6 \text{ V}$$

One *gigavolt* (GV) represents a billion (thousand-million) volts, so

$$1 \text{ GV} = 1,000,000,000 \text{ V}$$
$$= 10^9 \text{ V}$$

For Current

You can apply prefix multipliers to current values, just as you do with voltages. Often, you'll want to express current in terms of *milliamperes* (mA), where

$$1 \text{ mA} = 0.001 \text{ A}$$
$$= 10^{-3} \text{ A}$$

You'll also hear or read about *microamperes* (μA), where

$$1 \text{ μA} = 0.000001 \text{ A}$$
$$= 10^{-6} \text{ A}$$

Once in awhile, you'll encounter *nanoamperes* (nA), where

$$1 \text{ nA} = 0.000000001 \text{ A}$$
$$= 10^{-9} \text{ A}$$

For Resistance

With resistance, you'll frequently read or hear about large multiples of the ohm, but rarely about small fractions of an ohm. You can express resistance values in *kilohms* (k), where

$$1 \text{ k} = 1000 \text{ ohms}$$
$$= 10^3 \text{ ohms}$$

or in *megohms* (M), where

$$1 \text{ M} = 1,000,000 \text{ ohms}$$
$$= 10^6 \text{ ohms}$$

Once in awhile you'll come across a resistance expressed in *gigohms* (G), where

$$1 \text{ G} = 1,000,000,000 \text{ ohms}$$
$$= 10^9 \text{ ohms}$$

PROBLEM 3-1

Suppose that an electrical component carries 3 µA of direct current. What's this current in amperes?

✔ SOLUTION

From the table, observe that the prefix "micro-" (µ) represents fractional values equal to factors of 0.000001 (millionths). Therefore,

$$3 \mu A = 0.000003 \text{ A}$$

PROBLEM 3-2

The instruction manual for an old-fashioned vacuum-tube type power amplifier tells you that the components require a DC power supply capable of providing 2.2 kV for proper operation. What's this value in volts?

✔ SOLUTION

The table says that the prefix "kilo-" (k) stands for multiples of 1000 (thousands). Therefore

$$2.2 \text{ kV} = 2.2 \times 1000$$
$$= 2200 \text{ V}$$

PROBLEM 3-3

A resistor is rated at 47 M. What's this value in ohms?

✔ SOLUTION

From the table, you can see that the prefix "mega-" (M) stands for multiples of 1,000,000 (millions). Therefore

$$47 \text{ M} = 47 \times 1,000,000$$
$$= 47,000,000 \text{ ohms}$$

Ohm's Law

In a typical DC circuit, current flows through components having a certain amount of resistance. A certain EMF "pushes" the current through the resistive media. The current, voltage, and resistance interact in a predictable way. *Ohm's Law* defines this relationship in mathematical terms.

Three Ohm's Law Formulas

Figure 3-1 shows a DC circuit containing a source of DC voltage, a resistor, and an ammeter connected to measure the current in the circuit. Let E stand for the DC source voltage (in volts), let I stand for the current through the resistor (in amperes), and let R stand for the value of the resistor (in ohms). Three formulas express Ohm's law, as follows:

$$E = I R$$

$$I = E/R$$

$$R = E/I$$

TIP *If you want the above formulas to work right, you should use the standard electrical units: volts (V), amperes (A), and ohms. If quantities are specified in "prefix-multiplied" units other than volts, amperes, and ohms, you must convert to the standard units and then calculate. After that, you can express the resultant quantity (which will always turn out in volts, amperes, or ohms) into a unit with whatever multiplier you like.*

Some Calculations with Ohm's Law

To determine the current in a DC circuit, you must know the voltage and the resistance. You can also use Ohm's Law to calculate an unknown voltage when you know the current and the resistance, or to find an unknown resistance when you know the voltage and current.

For simplicity, assume that the wires in the circuit of Fig. 3-1 conduct perfectly, so that they have zero resistance. That way, you don't have to worry about their effect in the circuit. (In real life, wires always have a little bit of resistance, which can sometimes influence circuit behavior.)

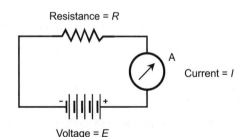

FIGURE 3-1 • A circuit for demonstrating current, voltage, and resistance calculations. Illustration for Problems 3-4 through 3-9.

PROBLEM 3-4

Suppose that the DC source in the circuit of Fig. 3-1 produces 6 V, and the resistor has a value of 2 ohms. What's the current?

SOLUTION

Plug in the numbers to the Ohm's Law formula for current to obtain

$$I = E/R$$
$$= 6/2$$
$$= 3 \text{ A}$$

PROBLEM 3-5

Suppose that the DC source in the circuit of Fig. 3-1 produces 12.0 V and the resistance equals 48.0 k. What is the current in amperes? In milliamperes? In microamperes?

SOLUTION

First, convert the resistance to ohms, getting

$$48.0 \text{ k} = 48.0 \times 1000$$
$$= 48{,}000 \text{ ohms}$$

Then plug the values directly into the formula for current in terms of voltage and resistance, obtaining

$$I = E/R$$
$$= 12.0/48{,}000$$
$$= 0.000250 \text{ A}$$

That's 0.250 thousandths of an ampere, so you can express it as 0.250 mA. It's also 250 millionths of an ampere, which you can express as 250 μA.

The Rule of Significant Figures

When they make calculations, competent engineers and scientists always follow the *rule of significant figures*, also called the *rule of significant digits*. After completing a calculation, you should round the final answer off so that it contains the *least* number of digits given in the input data.

As you follow this rule in Problem 3-5, you round off your final result to three significant digits, getting 0.000250 A, 0.250 mA, or 250 μA because you know the input values for voltage and resistance to an accuracy of three significant digits. If the voltage were given as 12.00 V and the resistance were given as 48.00 k (both accurate to four significant digits), then you would express the answer as 0.0002500 A, 0.2500 mA, or 250.0 μA. If the resistance were given as 48.000 k (to five significant digits) and the voltage were given as 12.0 V (to three significant digits), then you could state the answer to only three significant digits, getting 0.000250 A, 0.250 mA, or 250 μA.

When you see or use prefix-multiplier values or theoretical constants, you can consider them accurate to as many significant digits as you want. For example, the prefix "milli-" can mean a fractional quantity of 0.001 or 0.0010 or 0.00100 or 0.0010000000. Similarly, if you encounter a precise constant such as 2 or 10 or π (the mathematical constant *pi*, representing the ratio of a circle's circumference to its diameter) in a formula, you can express that constant to as many significant digits as you need. For example, you might round off π in a formula to get 3.14 or 3.1416 or 3.14159265359, depending on the accuracy of your data (such as measured current, voltage, or resistance).

PROBLEM 3-6

Suppose that the resistor in the circuit of Fig. 3-1 has a value of 80 ohms, and the measured current equals 1.2 μA. What's the potential difference across the resistor in volts? In microvolts?

✔ SOLUTION

Use the formula $E = I R$. First, convert the current to amperes, getting

$$1.2 \text{ μA} = 1.2 \times 0.000001$$
$$= 0.0000012 \text{ A}$$

Then plug in the numbers to obtain

$$E = I R$$
$$= 0.0000012 \times 80$$
$$= 0.000096 \text{ V}$$

That's 96 millionths of a volt, which equals 96 μV.

PROBLEM 3-7

Suppose that the resistor in the circuit of Fig. 3-1 has a value of 12.0 M, and the measured current equals 50.0 mA. What's the potential difference across the resistor in volts? In kilovolts?

✔ SOLUTION

Use the formula for voltage in terms of current and resistance. First, convert the resistance to ohms, obtaining

$$12.0\ M = 12.0 \times 1,000,000$$
$$= 12,000,000\ \text{ohms}$$

Then convert the current to amperes, getting

$$50.0\ mA = 50.0 \times 0.001$$
$$= 0.0500\ A$$

Finally, plug in the numbers to get

$$E = IR$$
$$= 0.0500 \times 12,000,000$$
$$= 600,000\ V$$

That's 600 × 1000 V, which equals 600 kV.

PROBLEM 3-8

If the voltage in the circuit of Fig. 3-1 equals 24 V and the ammeter shows 600 mA, what's the value of the resistor?

✔ SOLUTION

Use the Ohm's Law formula for resistance in terms of voltage and current. First, convert the current to amperes, getting

$$600\ mA = 600 \times 0.001$$
$$= 0.600\ A$$

Now you can plug in the values directly to obtain

$$R = E/I$$
$$= 24/0.600$$
$$= 40\ \text{ohms}$$

PROBLEM 3-9

If the DC source in the circuit of Fig. 3-1 provides 57 V and only 10 μA flow through the resistor, what's its resistance in ohms? In megohms?

SOLUTION

As you did in the solution to Problem 3-8, use the Ohm's Law formula for resistance in terms of voltage and current. First, convert the current to amperes, getting

$$10 \, \mu A = 10 \times 0.000001$$
$$= 0.000010 \, A$$

Next, plug in the values directly to get

$$R = E/I$$
$$= 57/0.000010$$
$$= 5,700,000 \text{ ohms}$$
$$= 5.7 \text{ M}$$

Power

In electricity, the term *power* refers to the rate at which a component, device, or system consumes energy. The standard unit of power is the *watt*, abbreviated as the uppercase, nonitalic letter W. When we express power as a variable in an equation, we use the uppercase, italic letter *P*.

Nonstandard Units of Power

Engineers often resort to "prefix-multiplied" units to express electrical power. Most often, you'll see the *kilowatt* (kW) or the *megawatt* (MW), where

$$1 \text{ kW} = 1000 \text{ W}$$
$$= 10^3 \text{ W}$$

and

$$1 \text{ MW} = 1,000,000 \text{ W}$$
$$= 10^6 \text{ W}$$

You'll also hear or read about the *milliwatt* (mW), the *microwatt* (μW), and the *nanowatt* (nW), where

$$1 \text{ mW} = 0.001 \text{ W}$$
$$= 10^{-3} \text{ W}$$

$$1 \text{ μW} = 0.000001 \text{ W}$$
$$= 10^{-6} \text{ W}$$

$$1 \text{ nW} = 0.000000001 \text{ W}$$
$$= 10^{-9} \text{ W}$$

Calculating DC Power

Figure 3-2 illustrates a circuit comprising a DC source, a motor, a voltmeter that measures the potential difference across the source and the motor, and an ammeter that measures the current through the motor. Let E stand for the voltmeter reading (in volts), let I stand for the ammeter reading (in amperes), and let R stand for the resistance of the motor (in ohms). You can use any of the following formulas to determine the power P (in watts) that the motor consumes. If you know the voltage and the current, then

$$P = E I$$

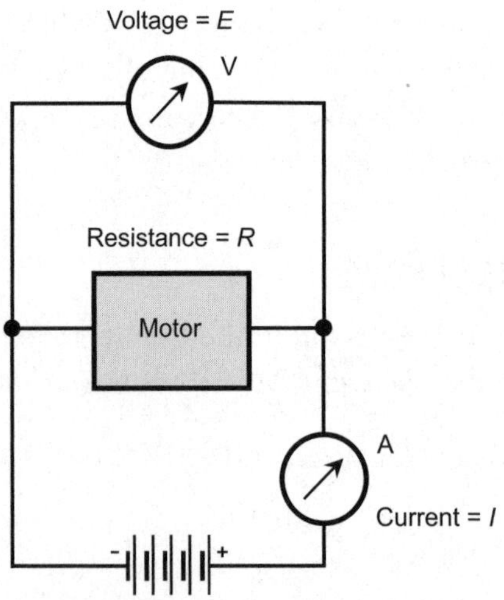

FIGURE 3-2 · A DC circuit for demonstrating power calculations. Illustration for Problems 3-10 through 3-15.

If you know the voltage and the resistance, then

$$P = E^2/R$$

If you know the current and the resistance, then

$$P = I^2 R$$

TIP *You should always work with units of volts (V), amperes (A), and ohms if you want your power calculations to come out in watts. If quantities are given in units other than volts, amperes, and ohms, you must convert to these units before you begin doing any arithmetic.*

In the following problems and calculations, assume that the wires, the DC source, and the ammeter in the circuit conduct perfectly, so that the only resistance in the circuit appears inside the motor. (An ideal voltmeter doesn't conduct any current, so it acts like an open circuit.)

PROBLEM 3-10

Suppose that the voltmeter in the circuit of Fig. 3-2 reads 5.6 V, while the ammeter reads 2.5 A. How much power does the motor consume? What's the DC source voltage?

 SOLUTION

Plug the numbers into the equation for power in terms of voltage and current to obtain

$$P = EI$$
$$= 5.6 \times 2.5$$
$$= 14\,W$$

The DC source voltage equals 5.6 V. You know this fact because the voltmeter is connected directly across the DC source, as well as directly across the motor.

PROBLEM 3-11

Imagine that, in the circuit of Fig. 3-2, you double the voltage across the motor to 11.2 V. (You can do that by doubling the DC source voltage.) But suppose that, despite the voltage increase, the current through the motor remains unchanged at 2.5 A. What happens to the power consumed by the motor?

SOLUTION

The motor consumes twice as much power as it did before. You double the E factor in the power equation while leaving the I factor constant, so you end up doubling the product of E and I, which equals P. You can demonstrate this particular case by plugging the new values into the formula for power in terms of voltage and resistance to get

$$P = EI$$
$$= 11.2 \times 2.5$$
$$= 28\ W$$

That's twice the power value of 14 W that you got when you solved Problem 3-10.

PROBLEM 3-12

Imagine that the DC source in the circuit of Fig. 3-2 provides the motor with 48.0 V, while the motor resistance equals 7.20 k. How much power does the motor consume in watts? In milliwatts?

SOLUTION

First, convert the resistance to ohms, obtaining $R = 7200$ ohms. Then plug the values into the formula for power in terms of voltage and resistance, getting

$$P = E^2/R$$
$$= 48.0 \times 48.0/7200$$
$$= 2304/7200$$
$$= 0.320\ W$$
$$= 320\ mW$$

PROBLEM 3-13

Suppose that you double the DC source voltage in the circuit of Fig. 3-2, while the motor resistance remains unchanged. What happens to the consumed power?

 SOLUTION

The consumed power will quadruple. You can verify this fact by examining the formula for power in terms of voltage and resistance. The formula tells you to square the voltage, so doubling it will increase the power by a factor of 2 squared, or 4. If you double the voltage from 48.0 V to 96.0 V in the situation of Problem 3-12 while leaving the resistance at 7200 ohms, you get

$$P = E^2/R$$
$$= 96.0 \times 96.0/7200$$
$$= 9216/7200$$
$$= 1.28\ W$$

That's four times the previous answer of 0.320 W.

PROBLEM 3-14

Suppose that the motor in the circuit of Fig. 3-2 has an internal resistance of 10.0 ohms, and the measured current through it equals 600 mA. How much power does the motor consume?

SOLUTION

Use the formula for power in terms of current and resistance. First, convert the current to amperes: 600 mA = 0.600 A. Then plug in the numbers, getting

$$P = I^2 R$$
$$= 0.600 \times 0.600 \times 10.0$$
$$= 0.360 \times 10.0$$
$$= 3.60\ W$$

PROBLEM 3-15

Suppose that the motor current decreases by a factor of 5 from its value in the scenario of Problem 3-14 and Fig. 3-2, while its resistance remains unchanged. What happens to the consumed power?

SOLUTION

The consumed power will go down to 1/25 of its previous value. Look at the formula for power in terms of current and resistance. You must square the current, so reducing it by a factor of 5 will decrease the final answer by a factor of 5 squared, or 25. If you change the current from 0.600 A to

0.120 A in the situation of Problem 3-14 (that's 1/5 as much current) while leaving the resistance at 10.0 ohms, you obtain

$$P = I^2 R$$
$$= 0.120 \times 0.120 \times 10.0$$
$$= 0.0144 \times 10.0$$
$$= 0.144\,W$$
$$= 144\,mW$$

That's 1/25 of the previous value of 3.60 W.

Energy

We can define electrical energy as power consumed over a specific period of time. The standard unit of energy is the *joule*, abbreviated as an uppercase, nonitalic letter J. Theoretically, a joule equals a *watt-second* (Ws), which represents the equivalent of one watt (1 W) expended for one second (1 s) of time. When we want to express energy as a variable in equations, let's symbolize it by writing an uppercase, italic letter Q. (We can't use E, which we've already used to represent voltage as a variable in equations.)

Energy Units

In real-world electrical circuits and systems, the joule constitutes a tiny unit. More often, you'll find electrical energy expressed in *watt-hours* (Wh). A watt-hour represents 1 W consumed for one hour (1 h) or 3600 seconds (3600 s), working out as 3.6 *kilojoules* (kJ). Therefore

$$1\,Wh = 3600\,Ws$$
$$= 3600\,J$$
$$= 3.6\,kJ$$

In medium-sized electrical systems, such as households, even the watt-hour is a small unit of energy. In situations like that, we can use the *kilowatt-hour* (kWh), which is the equivalent of 1 kW (1000 W) expended for 1 h, a quantity that works out as 3.6 *megajoules* (MJ). Mathematically,

$$1\,kWh = 1000\,Wh$$
$$= 1000 \times 3600$$
$$= 3,600,000\,J$$
$$= 3.6\,MJ$$

In large electrical networks such as the utility grids in cities, even the kilowatt-hour is a small unit! Then the *megawatt-hour* (MWh) is used—the equivalent of 1 MW (1,000,000 W) expended for 1 h. That works out as 3.6 *gigajoules* (GJ), so

$$1 \text{ MWh} = 1,000,000 \text{ Wh}$$
$$= 1,000,000 \times 3600$$
$$= 3,600,000,000 \text{ J}$$
$$= 3.6 \text{ GJ}$$

Three Energy Formulas

Figure 3-3 shows a circuit containing a DC source, an incandescent lamp, an ammeter to measure the current through the lamp, a switch to turn the lamp on and off, a timer to measure the length of time that the lamp stays illuminated, and a voltmeter to measure the EMF supplied by the DC source.

Assume that the wires and the timer have no resistance, so that they won't influence the outcome of your calculations, and so that you can have confidence that the voltmeter will indicate the true voltage across the lamp as well as the voltage from the DC source. Let E stand for the voltmeter reading (in volts), let I stand for the ammeter reading (in amperes), let R stand for the resistance of the lamp (in ohms), and let t stand for the timer reading (in hours).

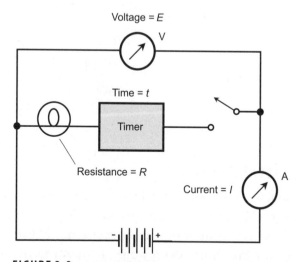

FIGURE 3-3 • A DC circuit for demonstrating energy calculations. Illustration for Problems 3-16 through 3-18.

You can use any of the following three formulas to determine the energy, Q (in watt-hours) consumed by the bulb:

$$Q = E I t$$

$$Q = E^2 t/R$$

$$Q = I^2 R t$$

TIP   *You should use units of volts (V), amperes (A), ohms, and hours (h) in order to ensure that your energy calculations will turn out in watt-hours. If quantities appear in units other than the standard ones, you must convert to the standard units before you input any numbers into the above formulas.*

PROBLEM 3-16

Suppose that in the circuit of Fig. 3-3, you close the switch, leave it closed for 2.0 h, and then open it again. While the switch is closed, the voltmeter indicates 12 V and the ammeter shows 1.5 A. How much energy, in watt-hours, does the bulb consume during this time period?

SOLUTION

Use the formula for energy in terms of voltage, current, and time. You're given the quantities in standard units, so you can plug in the numbers directly. Here, $E = 12$, $I = 1.5$, and $t = 2.0$, so

$$Q = EIt$$

$$= 12 \times 1.5 \times 2.0$$

$$= 36\,Wh$$

PROBLEM 3-17

Imagine that you don't know the ammeter reading in the circuit of Fig. 3-3, but you know that the switch remains closed for 30 minutes (30 min) and the voltmeter shows 6.00 V. Suppose that the resistance of the bulb equals 6.0 ohms when it glows. How much energy does the bulb consume during this period of time?

✔ **SOLUTION**

First, convert the time to hours: 30 min = 0.50 h. Therefore, $t = 0.50$. Also, note that $E = 6.00$ and $R = 6.0$. Use the formula for energy in terms of voltage, resistance, and time to obtain

$$Q = E^2 t / R$$
$$= 6.00 \times 6.00 \times 0.50/6.0$$
$$= 3.0 \text{ Wh}$$

PROBLEM 3-18

Suppose that, in the system shown by Fig. 3-3, you don't know the DC source voltage. However, you know that the lamp has a resistance of 10 ohms when it glows. You close the switch for 90 min, and during this time the ammeter reads 400 mA. How much energy does the bulb consume during this time period?

✔ **SOLUTION**

First, convert the time to hours: 90 min = 1.5 h. Therefore, $t = 1.5$. Then convert the current to amperes: 400 mA = 0.400 A. Therefore, $I = 0.400$. Note that $R = 10$. Plug these numbers into the formula for energy in terms of current, resistance, and time, getting

$$Q = I^2 R t$$
$$= 0.400 \times 0.400 \times 10 \times 1.5$$
$$= 2.4 \text{ Wh}$$

?

Still Struggling

Energy represents the equivalent of power multiplied by time, and power represents the equivalent of energy per unit of time. That's the essential qualitative difference between these two parameters in mathematical terms. If you imagine driving a car along a highway, then energy behaves like the total distance that you travel over a given period of time, and power acts like the speed at which you move at any instant in time. Power consumption can rise and fall from moment to moment, but energy consumption increases relentlessly as time passes.

QUIZ

This is an "open book" quiz. You may refer to the text in this chapter. You'll find the correct answers listed in the back of the book.

1. Figure 3-4 shows a circuit containing a source of variable DC voltage, a potentiometer (variable resistor), an ammeter to measure the current through the potentiometer, a switch that allows us to place the potentiometer in or out of the circuit at will, a timer to measure the length of time that the potentiometer carries current while the switch remains closed, and a voltmeter to measure the EMF supplied by the voltage source. Suppose that $E = 9.00$ V and $R = 90.0$ ohms. We close the switch. The ammeter tells us that

   A. $I = 10.0$ A.
   B. $I = 810$ A.
   C. $I = 100$ mA.
   D. $I =$ zero.

2. Suppose that we leave the switch closed in the system of Fig. 3-4 and Question 1, and then we adjust the potentiometer so that its resistance doubles compared with its previous value. If the source voltage remains the same, what happens to the current displayed by the ammeter?

   A. It drops to 1/4 of its previous value.
   B. It drops to half of its previous value.
   C. It does not change.
   D. All of the above! The current was zero before, and it's zero now.

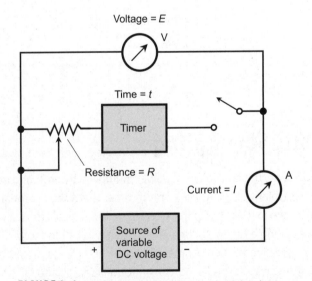

FIGURE 3-4 · Illustration for Quiz Questions 1 through 10.

3. Suppose that we connect a 13.5-V battery as the DC voltage source in the system of Fig. 3-4. We close the switch and read the ammeter, which tells us that $I = 250$ mA. Based on this information, we know that the potentiometer resistance is

A. $R = 54.0$ ohms.
B. $R = 3.38$ ohms.
C. $R = 0.844$ ohms.
D. $R = 0.0185$ ohms.

4. Suppose that we leave the switch closed in the system of Fig. 3-4 and Question 3, and then we adjust the potentiometer so that the ammeter displays a new current equal to three times its previous value. From this information, we know that the potentiometer's resistance has

A. increased to nine times its previous value.
B. increased to three times its previous value.
C. diminished to 1/3 of its previous value.
D. diminished to 1/9 of its previous value.

5. Suppose that we adjust the potentiometer to a resistance of 1.25 k in the system of Fig. 3-4. Then we close the switch and read the ammeter. It indicates that $I = 700$ μA (microamperes, not milliamperes). From this information, we know that the DC source provides

A. $E = 0.560$ nV.
B. $E = 1.79$ MV.
C. $E = 0.613$ μV.
D. $E = 875$ mV.

6. Suppose that we leave the switch closed in the system of Fig. 3-4 and Question 5, and then we double the DC source voltage and simultaneously reduce the potentiometer's resistance to half its previous value. The current that the ammeter displays will

A. remain unchanged.
B. increase to twice its previous value.
C. increase to four times its previous value.
D. diminish to half of its previous value.

7. Suppose that we connect a 32.0-V battery as the DC voltage source in the system of Fig. 3-4. We close the switch and read the ammeter, which tells us that $I = 25.0$ mA. Based on this information, we know that the potentiometer dissipates

A. 800 mW of power.
B. 20.0 mW of power.
C. 781 μW of power.
D. 195 nW of power.

8. Suppose that we leave the switch closed in the system of Fig. 3-4 and Question 7, and then we double the DC source voltage and simultaneously increase the

potentiometer's resistance to four times its previous value. The power that the potentiometer dissipates will

A. remain unchanged.
B. increase to twice its previous value.
C. increase to four times its previous value.
D. diminish to half of its previous value.

9. Suppose that we connect a 12-V battery as the DC voltage source in the system of Fig. 3-4. We close the switch and read the ammeter, which tells us that $I = 100$ mA. We leave the switch closed for 30 minutes. Based on this information, we know that during this period of time, the potentiometer dissipates

A. 0.12 Wh of energy.
B. 0.60 Wh of energy.
C. 12 Wh of energy.
D. 18 Wh of energy.

10. Suppose that we repeat the experiment described in Question 9, but with a 24.0-V battery instead of a 12-V battery. We don't change the potentiometer's resistance. In 30 minutes, the potentiometer will dissipate

A. 12 Wh of energy.
B. 9.6 Wh of energy.
C. 4.8 Wh of energy.
D. 2.4 Wh of energy.

chapter 4

Simple DC Circuits

A *DC circuit*, also called a *DC network*, comprises an interconnected group of electrical components that operates from a source of DC voltage. In this chapter, we'll see how voltage, current, and resistance interact in elementary DC circuits.

CHAPTER OBJECTIVES

In this chapter, you will

- Compare series and parallel circuit configurations.
- Learn how components and parameters interact in series-connected circuits.
- Learn how components and parameters interact in parallel-connected circuits.
- Convert resistance values to conductance values, and vice-versa.
- Apply Kirchhoff's Current Law and Kirchhoff's Voltage Law.

Series Circuits

To make a *series circuit*, we connect individual components end to end like the links of a chain. Figure 4-1 illustrates a series connection in block-diagram form, where several components (marked C) appear in series with a single source of DC voltage.

Voltage Sources in Series

When we connect two or more DC voltage sources in series, their voltages add up, as long as we make certain that the individual source polarities all go in the same direction. Imagine *n* batteries connected in series (where *n* represents a whole number), all connected "minus-to-plus," as shown in Fig. 4-2. Let E_1, E_2, E_3, ..., up to E_n represent the EMFs of the batteries, all expressed in volts. In theory, the total voltage E equals the sum of the EMFs of the individual batteries. Mathematically, we can write

$$E = E_1 + E_2 + E_3 + ... + E_n$$

In real-world networks, this equation constitutes an oversimplification. All voltage sources have a small amount of *internal resistance* that makes the actual net voltage across a series combination of sources slightly lower than the sum

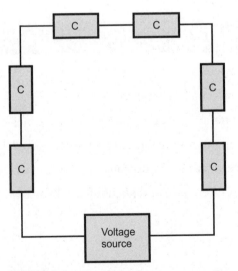

FIGURE 4-1 · In a series circuit, we connect the components (small boxes labeled C) end to end like the links of a chain

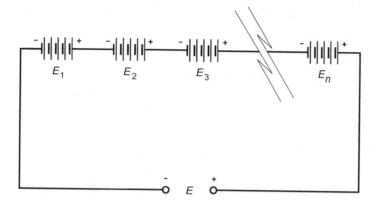

FIGURE 4-2 · When we connect sources of DC voltage in series, the total voltage equals the sum of the individual source voltages, assuming all the polarities agree.

of the individual source voltages. The discrepancy between the theoretical ideal (as the above formula would dictate) and the real-world voltage increases as external circuits demand more current. Nevertheless, you should remember the above formula as a basic principle.

TIP *If you connect several "fresh" batteries in series and test them with a voltmeter, first individually and then as a combination, you'll find that the foregoing ideal-ized formula "predicts" the net voltage within the limits of experimental error.*

Polarity Reversal

Now imagine that you turn one or more of the batteries in the configuration of Fig. 4-2 around and connect it "backwards." In that case, you must subtract, rather than add, the voltage of each "backward" battery when you calculate the net voltage produced by the series combination.

Suppose, for example, that you reverse the polarity of battery number 2 (with voltage E_2). Figure 4-3 shows the new situation, in which the series-combination voltage E turns out as

$$E = E_1 - E_2 + E_3 + ... + E_n$$

Resistances in Series

When we connect two or more *resistors* (components having resistance) in series, their ohmic values simply add up. Imagine n resistors in series. Let R_1, R_2, R_3, ..., up to R_n represent their values, all expressed in ohms. The total resistance

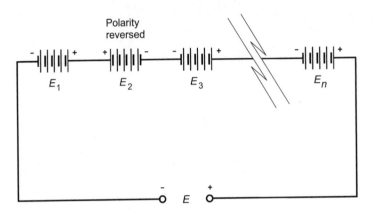

FIGURE 4-3 · A series combination of DC voltage sources in which one source has its polarity reversed. This voltage subtracts from, rather than adds to, the total.

R equals the sum of the resistances of the individual resistors (Fig. 4-4A). Mathematically, we can write this fact as

$$R = R_1 + R_2 + R_3 + \ldots + R_n$$

The above equation represents a theoretical ideal. It neglects the resistance of the wires that interconnect the resistors. In most real-world circuits of this kind, we can ignore the wire resistance. But if the ohmic values of the resistors are so low that they compare favorably with the resistance in the wire, the foregoing formula gives us an oversimplification. In that sort of scenario (Fig. 4-4B), we must treat the wire resistance R_w as another component in the formula, so that the total resistance becomes

$$R = R_1 + R_2 + R_3 + \ldots + R_n + R_w$$

TIP *Polarity has no meaning, and therefore, no significance, with ordinary resistors. In theory, resistors don't have poles. We can turn a resistor around, and our circuit will work exactly as it did before. The opposition to DC by a resistor in one direction always equals its opposition to DC in the other direction.*

Current Sources in "Series"

In a series DC circuit, only one branch exists in which current can flow. The current at any point, therefore, equals the current at any other point. Technically, it makes no sense to talk about "current sources connected in series" because every component in a series circuit carries the same current, however large or

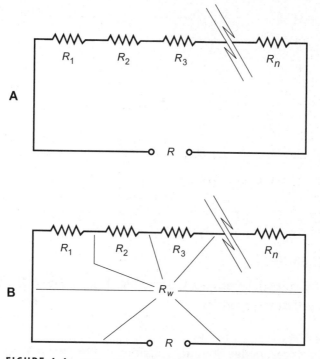

FIGURE 4-4 • At A, a series combination of resistors in which the wire resistance is negligible compared with the values of the resistors. At B, we must take the wire resistance (R_w) into account if the resistors have extremely small ohmic values.

small. If the current through any single component goes up or down, the current through every other component goes up or down along with it.

PROBLEM 4-1

What's the DC voltage E between the two terminals in Fig. 4-5? Pay special attention to the battery polarities.

SOLUTION

The batteries are all connected with the same polarity; the terminals all go "minus-to-plus" (as they should in any circuit intended for practical use!). Therefore, their voltages add up directly. You can calculate the net voltage as

$$E = 6 + 9 + 3 + 12$$
$$= 30\,\text{V}$$

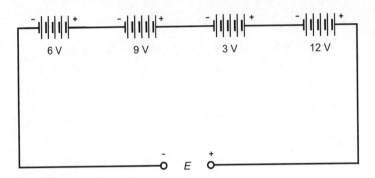

FIGURE 4-5 · Illustration for Problem 4-1.

PROBLEM 4-2

What's the DC voltage E between the two terminals in Fig. 4-6? Pay special attention to the battery polarities.

SOLUTION

This circuit resembles the one described in Problem 4-1, except that the battery second from the left has been "turned around" so that its voltage subtracts from, rather than adds to, the total. The voltage at the terminals is therefore

$$E = 6 + (-9) + 3 + 12$$

$$= 12\,V$$

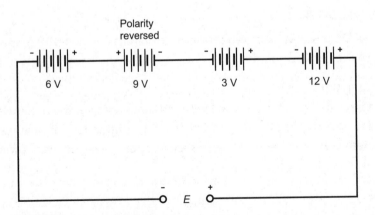

FIGURE 4-6 · Illustration for Problem 4-2.

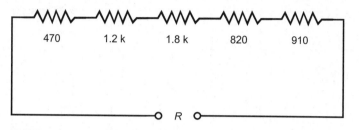

FIGURE 4-7 · Illlustration for Problem 4-3. Resistances are expressed in ohms, where k indicates kilohms (multiples of 1000 ohms).

TIP   *You can easily conduct an experiment to test the "real-world" outcome in a situation such as the one described in Problem 4-2 and its solution. In my book* Electricity Experiments You Can Do at Home *(McGraw-Hill, 2010), I discuss the results of such an experiment. You can use simple size AA flashlight cells and an inexpensive voltmeter.*

PROBLEM 4-3

What's the DC resistance *R* between the terminals in Fig. 4-7? Pay special attention to the resistance units. Express the answer in ohms, and also in kilohms. Assume that the wire has no resistance.

SOLUTION

Convert all values to ohms before doing any calculations. Note that 1.2 k = 1200 ohms and 1.8 k = 1800 ohms. Therefore

$$R = 470 + 1200 + 1800 + 820 + 910$$
$$= 5200 \text{ ohms}$$
$$= 5.2 \text{ k}$$

Technically, we can claim only two significant figures of accuracy here because two of the resistor values (1.2 k and 1.8 k) contain only two digits.

Parallel Circuits

To make a *parallel circuit*, we connect components together so that, in a schematic diagram, they appear something like the rungs of a ladder. Figure 4-8 shows a simple example. Some current flows through every branch, where each

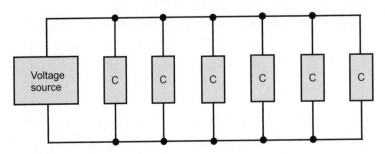

FIGURE 4-8 · In a parallel circuit, we connect the components (small boxes labeled C) across each other like the rungs of a ladder.

branch contains one component marked C. The branch currents aren't necessarily all the same, but every component receives the same voltage as every other (and also the same voltage as the source) because each component is connected directly across the voltage source.

Voltage Sources in Parallel

Normally, we won't find more than one voltage source in a parallel circuit. Sometimes we'll want to connect cells or batteries in parallel to get a voltage source that can deliver more *current* than a single cell or battery can do alone; but if we do that, we should make certain that all of the cells or batteries are identical. If the voltages differ, some of the cells or batteries will drive current through others, and that's no good!

In a parallel circuit, the poles of the voltage sources must all go "plus-to-plus" and "minus-to-minus." Otherwise, we'll end up with short-circuit loops containing pairs of voltage sources in series with no intervening resistance between them. That situation causes a massive waste of energy. It can also create a physical hazard because, in effect, it places a short circuit across both sources. To see a situation like this, draw a schematic diagram of two batteries connected in parallel, with their polarities going "minus-to-plus." You'll have a short-circuited "double battery."

The output voltage of a properly designed parallel combination of cells or batteries equals the voltage of any single one of them. The total deliverable current I of a parallel combination of n cells or batteries equals the sum of the deliverable currents, I_1, I_2, I_3, ..., and I_n, of each one (Fig. 4-9). We can denote this fact mathematically as

$$I = I_1 + I_2 + I_3 + ... + I_n$$

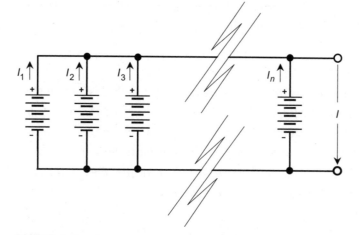

FIGURE 4-9 · When we connect two or more batteries having equal voltage in parallel with each other and with their polarities all in agreement, the total deliverable current equals the sum of the individual deliverable currents.

TIP *In Fig. 4-9, the arrows point in the direction of the conventional (theoretical) currents, which, as we learned in Chapter 2, flow away from the positive poles and toward the negative poles.*

Resistances in Parallel

When we connect two or more resistors in parallel (Fig. 4-10), their ohmic values combine in a rather complicated way. Consider n resistors in parallel. Let

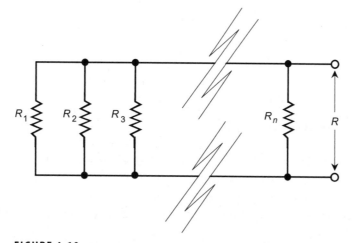

FIGURE 4-10 · Several DC resistances connected in parallel.

R_1, R_2, R_3, ..., and R_n represent their values, all expressed in ohms. If R represents the total resistance in ohms, then

$$R = 1/(1/R_1 + 1/R_2 + 1/R_3 + ... + 1/R_n)$$

TIP *The parallel-resistance formula works based on the assumption that the circuit wiring has no resistance. If the wire has significant resistance compared with the values of the resistors themselves, then some of the wire resistance adds in series with each resistor, and some of the wire resistance adds in series with the whole combination. In most practical circuits, we can neglect the wire resistance because the resistors usually have ohmic values that greatly exceed the wire resistance.*

The Siemens

The mathematical reciprocal of resistance is known as *conductance*. Engineers express electrical conductance using a unit called the *siemens*. The word "siemens" (which serves in the plural sense as well as the singular sense) is abbreviated as an uppercase, nonitalic letter S. We symbolize conductance as a variable in equations using the uppercase, italic letter G. If we let G represent the conductance of a particular component in siemens and R represent its resistance in ohms, then

$$G = 1/R$$

and

$$R = 1/G$$

Conductance values in parallel add up like resistance values do in series. If G_1, G_2, G_3, ..., and G_n represent the conductances, in siemens, of the individual resistors in a parallel combination, then the composite conductance G, also in siemens, is

$$G = G_1 + G_2 + G_3 + ... + G_n$$

? Still Struggling

Some people find the foregoing methodology easier to understand than the "resistance-only" formula when dealing with parallel circuits. However, if you try to calculate parallel resistances by converting them to conductances first, you must remember to convert the total conductance back to resistance at the end of the process.

Currents in Parallel Branches

Imagine a parallel circuit with n branches, connected across a DC voltage source such as a battery (Fig. 4-11). Let $I_1, I_2, I_3, ...,$ and I_n represent the currents in each of the branches, all expressed in amperes. The total current I drawn from the battery, also in amperes, equals the sum of the branch currents. Mathematically

$$I = I_1 + I_2 + I_3 + ... + I_n$$

PROBLEM 4-4

Suppose that we connect five lantern batteries in parallel. Each battery supplies 6.3 V. What's the voltage of the resulting combination?

SOLUTION

The voltage of a parallel set of batteries, all having the same voltage, equals the voltage of any single battery considered by itself. Therefore, the combination supplies 6.3 V.

PROBLEM 4-5

Suppose, in the preceding problem, that each battery can reliably provide up to 1.3 A of current. How much current can the parallel combination of batteries reliably provide?

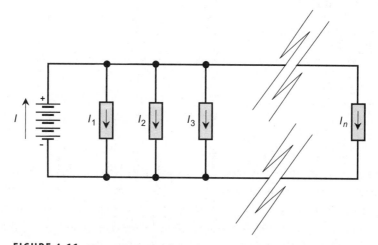

FIGURE 4-11 • The total current drawn by a parallel circuit equals the sum of the currents in the individual branches.

SOLUTION

In theory, the combination can provide a total current I equal to the sum of the individual deliverable currents. Therefore

$$I = 1.3 + 1.3 + 1.3 + 1.3 + 1.3$$

$$= 5 \times 1.3$$

$$= 6.5 \text{ A}$$

PROBLEM 4-6

What's the current I_3 in the circuit of Fig. 4-12?

 **SOLUTION**

You can calculate I_3 using the formula for the total current in a parallel combination to get an equation with a single unknown, and then solve with simple algebra. Start with the fact that

$$18 = 6 + 3 + I_3 + 5$$

where all values are expressed in amperes. Subtract the sum $(6 + 3 + 5)$ from both sides of this equation to get

$$18 - (6 + 3 + 5) = I_3$$

which simplifies to

$$18 - 14 = I_3$$

Therefore, $I_3 = 4$ A.

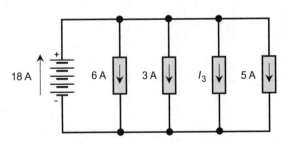

FIGURE 4-12 • Illustration for Problem 4-6.branches.

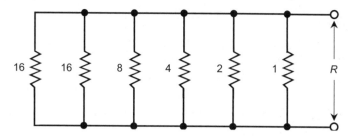

FIGURE 4-13 · Illustration for Problems 4-7 and 4-8. Resistances are expressed in ohms.

PROBLEM 4-7

What's the resistance R across the parallel combination of resistors shown in Fig. 4-13?

✔ SOLUTION

You can use the formula for parallel resistances to get

$$R = 1 / (1/16 + 1/16 + 1/8 + ¼ + ½ + 1)$$
$$= 1/2$$
$$= 0.5 \text{ ohms}$$

PROBLEM 4-8

In the scenario of the preceding problem, convert the resistor values to conductances, and then calculate the conductance G across the entire combination.

✔ SOLUTION

The individual conductance values, proceeding from left to right, are 1/16 S, 1/16 S, 1/8 S, ¼ S, ½ S, and 1 S. Adding up all these conductance values yields

$$G = 1/16 + 1/16 + 1/8 + ¼ + ½ + 1$$
$$= 2 \text{ S}$$

TIP *Note that the above-derived composite conductance in siemens equals the reciprocal of the composite resistance in ohms.*

Kirchhoff's Laws

Two of the most important DC network principles involve currents that flow into and out of specific circuit points, and the sums of the voltages around closed loops. These rules are often called *Kirchhoff's First Law* and *Kirchhoff's Second Law*. Some engineers call them *Kirchhoff's Current Law* and *Kirchhoff's Voltage Law*, respectively.

The Law for Current

The sum of the currents that flow into any point in a DC circuit equals the sum of the currents that flow out of that point. Figure 4-14 shows a simple example. In this case, the current entering point Z equals $I_1 + I_2$, while the current emerging from point Z equals $I_3 + I_4 + I_5$. Therefore

$$I_1 + I_2 = I_3 + I_4 + I_5$$

Kirchhoff's Current Law holds true no matter how many branches lead into or out of a branch point. When you make calculations using this rule, you must express all current values in the same units (amperes, milliamperes, microamperes, or whatever).

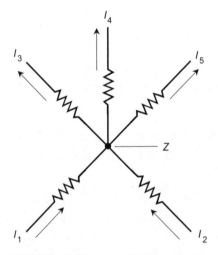

FIGURE 4-14 · Kirchhoff's Current Law. The total current entering point Z equals the total current emerging from point Z.

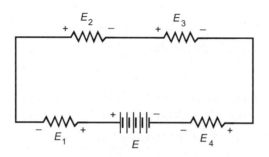

FIGURE 4-15 · Kirchhoff's Voltage Law. The sum of the voltages around the closed loop equals zero when we take the individual component polarities into account.

The Law for Voltage

The sum of all the voltages, as you go around any closed loop in a DC circuit from some fixed point and return there from the opposite direction, and taking polarity into account, always equals zero volts (0 V). In the situation of Fig. 4-15, for example, you know that

$$E_1 + E_2 + E_3 + E_4 + E = 0 \text{ V}$$

which you can also state as

$$E_1 + E_2 + E_3 + E_4 = -E$$

As with the law for current, you must specify all the potential differences in the same units (volts, kilovolts, millivolts, or whatever).

? Still Struggling

Kirchhoff's Current Law arises from the fact that electrical current can't appear from nowhere, and it can't vanish into nothingness. Think of water in a plumbing system, where all the water that goes into any point must come out of that point. Kirchhoff's Voltage Law arises from the fact that a potential difference can't exist between any single circuit point and itself. Also, a potential difference can't exist between any two points directly connected to each other with a perfect conductor.

PROBLEM 4-9

What's the current I_2 in the circuit of Fig. 4-16?

SOLUTION

Use Kirchhoff's Current Law. All the current values appear in amperes, so you don't have to do any unit conversions. You can determine the total current I_{out} coming out of the branch point as

$$I_{out} = 4 + 5 + 7$$
$$= 16 \text{ A}$$

The sum of the currents I_{in} entering the branch point must also equal 16 A. You know two of the branch currents going into the point, so determining the unknown current I_2 is a simple algebra problem. You can set up the equation

$$16 = 6 + I_2 + 4$$

where all units are expressed in amperes. It solves to $I_2 = 6$ A.

PROBLEM 4-10

Refer to Fig. 4-17. What's the battery voltage E?

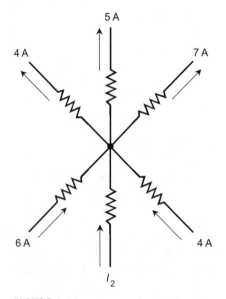

FIGURE 4-16 • Illustration for Problem 4-9.

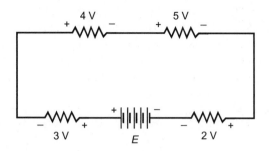

FIGURE 4-17 · Illustration for Problem 4-10.

SOLUTION

Use Kirchhoff's Voltage Law. The sum of all the voltages, taking polarity into account, must equal 0 V. Therefore

$$E + 3 + 4 + 5 + 2 = 0$$

where all units are expressed in volts. This equation solves algebraically to

$$E = -14\,V$$

TIP   *The foregoing answer comes out negative because, as the problem is set up, the potential differences across the resistors are all described as positive voltages. The polarity of the voltage across the battery opposes the polarities of the voltages across the resistors, so the battery voltage ends up with a minus sign.*

QUIZ

This is an "open book" quiz. You may refer to the text in this chapter. You'll find the correct answers listed in the back of the book.

1. Figure 4-18 shows four DC batteries connected in series. In theory, if we connect a voltmeter to the terminals marked E, we should see a reading of
   A. 15 V.
   B. 18 V.
   C. 21 V.
   D. 24 V.

2. Suppose that we reverse the polarity of the 3.0-V battery in the system shown by Fig. 4-18. If we connect a voltmeter to the pair of terminals marked E, we should see a reading of
   A. 15 V.
   B. 18 V.
   C. 21 V.
   D. 24 V.

3. Figure 4-19 shows a set of five resistors connected in a combination series/parallel circuit. The net resistance R equals
   A. 550 ohms.
   B. 600 ohms.
   C. 650 ohms.
   D. 750 ohms.

4. If we remove one of the 100-ohm resistors from the circuit of Fig. 4-19, leaving an open circuit in its place, then the net resistance R will equal
   A. 500 ohms.
   B. 550 ohms.
   C. 600 ohms.
   D. 650 ohms.

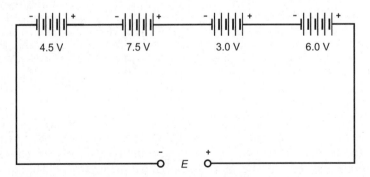

FIGURE 4-18 · Illlustration for Quiz Questions 1 and 2.

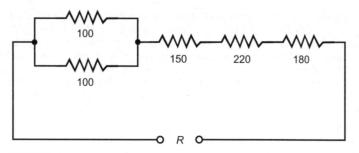

FIGURE 4-19 · Illustration for Quiz Questions 3 through 5. All resistance values are expressed in ohms.

5. If we short-circuit one of the 100-ohm resistors in the circuit of Fig. 4-19, then the net resistance R will equal

    A. 450 ohms.
    B. 500 ohms.
    C. 550 ohms.
    D. 600 ohms.

6. Figure 4-20 shows a network of six resistors. Suppose that the free ends of the top three resistors all go to the negative terminal of a battery, while the free ends of the bottom three resistors all go to the positive terminal of the same battery. How much conventional current flows out through the point marked by the query symbol (upper left), away from the resistor?

    A. 0.25 A
    B. 0.50 A
    C. 1.00 A
    D. The premise is wrong! Conventional current does not flow out of the point marked by the query symbol, away from the resistor. Instead, conventional current flows into that point, toward the resistor.

7. Suppose that, in the system described in Question 6 and illustrated by Fig. 4-20, we cut the battery voltage in half but don't change any of the resistance values. In this case, how much conventional current will flow out through the point marked by the query symbol away from the resistor? Here's a hint: All of the current values shown in Fig. 4-20 will change, not only the current value marked by the query symbol. Here's another hint: Although this problem might seem difficult, we can deduce the answer based on what we've learned about DC circuits.

    A. 0.25 A
    B. 0.50 A
    C. 1.00 A
    D. The premise is wrong. Conventional current won't flow out of the point marked by the query symbol, away from the resistor. Instead, conventional current will flow into that point, toward the resistor, and it will be half as great as before.

8. Suppose that, in the system described in Question 6 and illustrated by Fig. 4-20, we double all of the resistance values but don't change the battery voltage. In this case, how much conventional current will flow out through the point marked by the query symbol away from the resistor? Here's a hint: All of the current values shown in Fig. 4-20 will change, not only the current value marked by the query symbol. Here's another hint: Although this problem might seem difficult, we can deduce the answer based on what we've learned about DC circuits.

   A. 0.25 A
   B. 0.50 A
   C. 1.00 A
   D. The premise is wrong. Conventional current won't flow out of the point marked by the query symbol, away from the resistor. Instead, conventional current will flow into that point, toward the resistor, and it will be 1/4 as great as before.

9. Figure 4-21 shows a network of four resistors connected in series with a battery. How much potential difference appears across the resistor marked with the query symbol (upper right)?

   A. 2.5 V
   B. 3.3 V
   C. 5.0 V
   D. 10.0 V

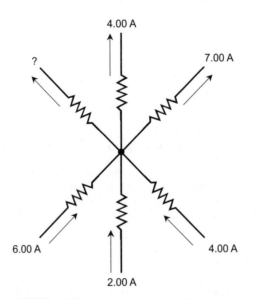

FIGURE 4-20 · Illustration for Quiz Questions 6 through 8.

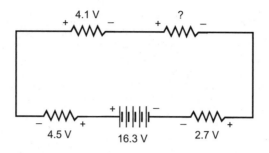

FIGURE 4-21 · Illustration for Quiz Questions 9 and 10.

10. Suppose that, in the system described in Question 9 and illustrated by Fig. 4-21, we double all of the resistance values but don't change the battery voltage. In this case, how much potential difference will appear across the resistor marked with the query symbol? Here's a hint: Although this problem might seem difficult, we can deduce the answer based on what we've learned about DC circuits.

    A. 2.5 V
    B. 3.3 V
    C. 5.0 V
    D. 10.0 V

chapter 5

Cells and Batteries

Specialized hardware components can generate electricity from chemical reactions, from radiant energy, or from combustible fuels. Let's see how they work.

CHAPTER OBJECTIVES

In this chapter, you will

- Get familiar with electrochemical-energy storage and generation methods.
- Distinguish between primary and secondary cells.
- See how multiple cells interconnect to form batteries.
- Learn how specialized cells convert radiant energy, especially visible light, into DC electricity.
- Look into the future of electrochemical power-generation technology.

Electrochemical Energy

An *electrochemical cell* converts *chemical energy* (a form of *potential energy*) into electrical energy. When we connect two or more electrochemical cells in series, we get an *electrochemical battery*. Electrochemical cells and batteries find widespread application in portable electronic equipment, communications satellites, and sources of emergency power.

A Basic Cell

Figure 5-1 illustrates the basic components of a *lead-acid cell*. An electrode made of lead and another electrode made of lead dioxide, immersed in a sulfuric-acid solution called the *electrolyte*, acquire a potential difference as the acid reacts with the metals. This voltage can drive current through a load. The amount of current that a lead-acid cell can reliably provide depends on the "freshness" of its acid, its mass and volume, as well as the combined surface areas of its plates. In a battery made from two or more identical lead-acid cells, the voltage depends on the number of series-connected cells.

If we leave a lead-acid cell connected to a constant load for a long time, the current through the load gradually decreases. The cell's electrodes *corrode*, acquiring a coating that interferes with the flow of current. The nature of the acid changes as well; it grows less "fresh" (in effect, less concentrated). Eventually, all of the chemical energy originally present in the acid is converted to electrical energy, the output current drops to zero, and a potential difference no longer exists between the two electrodes.

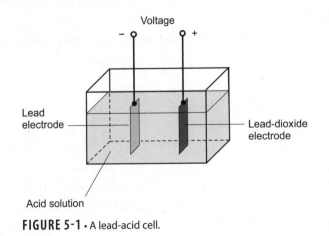

FIGURE 5-1 · A lead-acid cell.

Primary and Secondary Cells

Some electrochemical cells become useless, and must be thrown away, once their chemical energy has all changed to electricity and been consumed or dissipated. We call this type of cell a *primary cell*. Other kinds of cells, such as the lead-acid system described above, can get chemical energy back again by means of a controlled *recharging* process. We call such a component a *secondary cell*.

Most primary cells contain a chemical paste along with metal electrodes. These primary cells go by names such as *dry cell*, *zinc-carbon cell*, or *alkaline cell*. We'll find them in supermarkets and department stores. We can also find some types of secondary cells in everyday stores. These secondary cells cost more than ordinary dry cells, and they require a *recharging unit* that costs a few dollars. However, we can reuse them hundreds of times, so they pay for themselves and their recharging unit several times over.

An *automotive battery* consists of several secondary cells connected in series and placed in a single container. The combination recharges from the vehicle's *alternator* (a small electric generator driven by the vehicle's engine) or from an outside recharging unit. The individual cells resemble the one shown in Fig. 5-1. It's dangerous to short-circuit the terminals of such a battery, or even to subject it to an exceptionally heavy load (demand excessive current from it) because the extreme current can heat up the acid and cause it to boil out. Such a battery must always remain in a "right-side-up" position so that the acid doesn't leak or spill.

WARNING! *Never directly short-circuit any electrochemical cell or battery for more than a couple of seconds. It will likely overheat and might start a fire. In the extreme case, a short-circuited battery can actually explode! Automotive batteries present a special danger in this respect. You should never short-circuit an automotive battery, even for a fraction of a second.*

Standard Cell

Most cells produce a potential difference of 1.0 V to 1.8 V between their positive and negative electrodes. Some types of cells, known as *standard cells*, generate predictable and precise voltages under *no-load conditions* (when we don't demand significant current from them). An example is the *Weston standard cell*, which produces 1.018 V at room temperature. It has an electrolyte solution of cadmium sulfate. The positive electrode is made of mercury sulfate and elemental mercury, and the negative electrode comprises a mercury-and-cadmium mixture (Fig. 5-2).

TIP *When we operate a properly constructed Weston standard cell at room temperature, its voltage remains constant for a long period of time, so that we can use it as a reference source for calibrating laboratory voltmeters.*

Storage Capacity

Any cell or battery has a certain amount of *electrical storage capacity* that we can specify in watt-hours (Wh) or kilowatt-hours (kWh). Engineers sometimes specify a battery's capacity in terms of the mathematical *integral* of deliverable current with respect to time, in units of *ampere-hours* (Ah). The energy capacity in watt-hours equals the ampere-hour capacity multiplied by the battery voltage.

A battery with a rating of 20 Ah can provide 20 A for 1 h, or 1 A for 20 h, or 100 mA for 200 h. In fact, there exist an infinite number of current/time combinations, and almost any of them (except for the extremes) can be put to practical use. Engineers specify two fundamental parameters for electrochemical cells and batteries: the *shelf life* and the *maximum deliverable current*.

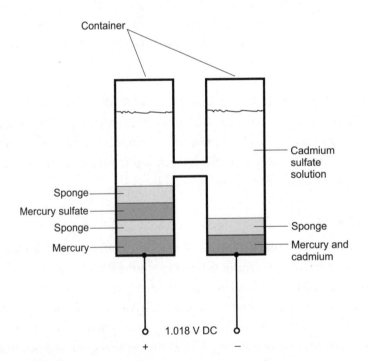

FIGURE 5-2 · A Weston standard cell.

We define the shelf life as the length of time that a cell or battery will remain usable if we never connect it to a load. Most common cells and batteries have shelf lives expressible in years. The maximum deliverable current equals the highest current that a cell or battery can drive through a load without the voltage dropping significantly because of the component's *internal resistance*, and without creating a fire or explosion hazard. The maximum deliverable current for a particular type of cell, when fully charged, depends on its mass and volume.

Polarity

A cell or battery always has a positive pole and a negative pole. As you've learned, theoretical (conventional) current, as defined by physicists, flows from the positive pole to the negative pole of a cell or battery in an electrical circuit, even though the individual electrons travel generally away from the negative pole and toward the positive pole.

In some electrical systems, DC polarity doesn't matter. In a lantern, for example, you can connect the battery in either direction, and the lamp will light up as long as electrical contact is properly maintained. If you turn all the cells around in a flashlight, and if the bulb still makes contact and all the cell contacts touch one another, the device will work exactly as it does with the cells installed "correctly." If you turn only one of the cells around in a two-cell flashlight, however, the bulb won't illuminate because the voltages from the cells conflict, in effect "canceling each other out."

In most electronic systems, in contrast with simple electrical devices, battery polarity is critical. You must make sure that you install the battery "the right way." Otherwise, the internal circuits won't work, and damage can occur to sensitive components. Portable electronic devices in which the battery polarity matters include radio receivers and transmitters, remote control boxes, cell phones, electric clocks, and notebook computers.

TIP *If you're not sure if the cell or battery polarity makes a difference for a particular device, assume that it matters! Such devices usually have small drawings or polarity signs in the "battery box" that tell you which way the cells or battery should go.*

Discharge Curves

When we use an *ideal cell* or *ideal battery*, it delivers a constant current for awhile, and then the current starts to decrease. Some types of cells and batteries approach ideal behavior, exhibiting a *flat discharge curve* (Fig. 5-3A). Other, not-so-ideal cells or batteries produce output current that decreases gradually from the beginning of use, giving us a *declining discharge curve* (Fig. 5-3B).

When the maximum deliverable current has decreased to about half of its initial value, we say that a cell or battery is "weak" or "low." At this time, we should replace it or recharge it before continuing to operate anything from it. If we allow a cell or battery to run down until the maximum deliverable current drops to nearly zero, we call the cell or battery "dead" or "depleted."

In either of the graphs in Fig. 5-3, the area under the curve (the area of the region bounded by the curve and the two coordinate axes) represents the storage capacity in units of current multiplied by time. For example, if we graduate the time coordinate in hours and the current coordinate in amperes, then the total area enclosed by the curve represents the storage capacity in ampere-hours.

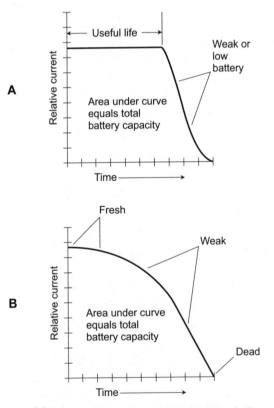

FIGURE 5-3 • At A, a flat discharge curve. At B, a declining discharge curve.

PROBLEM 5-1

Suppose that we connect a Weston standard cell to a voltmeter, and the meter produces the reading shown in Fig. 5-4. Based on this indication, we question the meter's accuracy! Approximately how far off is the meter reading?

✔ SOLUTION

A Weston standard cell produces 1.018 V. The meter indicates approximately 1.5 V. The meter reading appears almost 0.5 V too high, representing an error of nearly 50%.

PROBLEM 5-2

Suppose that the meter portrayed in Fig. 5-4 always reads 50% too high regardless of the applied voltage (as long as the voltage doesn't exceed the full-scale value). If we connect a flashlight cell that provides 1.50 V to the meter, what should we anticipate?

✔ SOLUTION

We should expect the meter needle to travel all the way to the right-hand end of the scale and "hit the pin." A reading 50% above 1.50 V equals 1.50 × 1.50 V, or 2.25 V. That voltage exceeds the full-scale range of the meter.

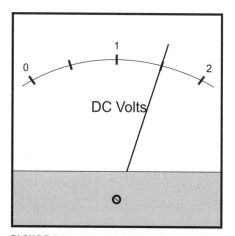

FIGURE 5-4 • Illustration for Problems 5-1 and 5-2.

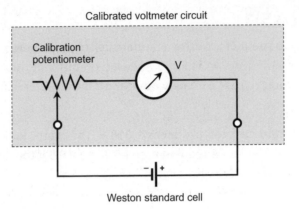

Calibrated voltmeter circuit

FIGURE 5-5 • Illustration for Problem 5-3.

PROBLEM 5-3

How can we modify the above-described voltmeter to make it produce accurate readings?

✔ SOLUTION

We can connect the meter in series with a potentiometer having an extremely large value (several megohms), as shown in Fig. 5-5. Then we can connect a Weston standard cell across the combination. Once we've done that, we can adjust the potentiometer until the meter reading slightly exceeds 1 V (the needle comes to rest "a hair's breadth" to the right of the hash mark for 1 V). This action should correct the error over the entire range of the meter, as long as we use the circuit within the dashed box (and not only the meter itself) to measure DC voltages that never exceed 2 V.

Common Cells and Batteries

The electrochemical cells sold in stores and used in convenience items, like flashlights and transistor radios, produce approximately 1.5 V. You can find them in sizes called AAA (very small), AA (small), C (medium), and D (large). Batteries made from combinations of electrochemical cells are usually rated at 6 V (four cells in series) or 9 V (six cells in series).

Zinc-Carbon

Zinc-carbon cells have fairly long shelf life. The zinc forms the outer case, and also constitutes the negative electrode. A carbon rod serves as the positive electrode. The electrolyte is a paste of manganese dioxide and carbon. Zinc-carbon cells don't cost much money. They work best at moderate temperatures, and in applications where the current drain ranges from moderate to high. They don't function very well in extremely cold environments (temperatures below the freezing point of water).

Alkaline

Alkaline cells have granular zinc for the negative electrode, potassium hydroxide as the electrolyte, and a substance called a *polarizer* as the positive electrode. An alkaline cell can work at lower temperatures than a zinc-carbon cell can. You can expect an alkaline cell or battery to last a long time in a low-current electronic device, such as a portable calculator or electronic clock. Alkaline cells and batteries last far longer on the shelf than zinc-carbon cells or batteries do, but they cost more.

Transistor

Transistor batteries are box-shaped with dimensions of approximately $5 \times 15 \times 25$ mm, equipped with clamp-on terminals. They provide 9 V and comprise six tiny zinc-carbon or alkaline cells in series. Transistor batteries are used in very-low-current electronic devices that function on an intermittent (as opposed to continuous-duty) basis, such as wireless garage-door openers, television and stereo remote-control boxes, smoke detectors, carbon-monoxide detectors, and electronic calculators.

Lantern

Lantern batteries have considerable bulk and mass, and can deliver a fair amount of current. This type of battery usually consists of four zinc-carbon or alkaline cells in series, producing about 6 V DC. One type of lantern battery has spring contacts on the top. The other type has thumbscrew terminals. A single lantern battery can keep a small, low-voltage DC incandescent bulb lit for quite a long time. A pair of them in series can deliver enough power to operate a small portable or mobile radio transceiver (such as a citizens-band or amateur radio set) for a few hours.

Silver-Oxide

Silver-oxide cells are usually molded into button-like shapes. For this reason, they're sometimes called *button cells* (although some other types of cells also have this shape). Silver-oxide cells can fit inside wristwatches, subminiature calculators, and small cameras. They come in various sizes and thicknesses, supply 1.5 V, and offer excellent energy storage capacity per unit of volume and mass. This type of cell is known for exhibiting a flat discharge curve, so you can expect it to work okay all the way up until it "dies." Silver-oxide cells can be stacked to make compact batteries.

Mercury

Mercury cells, also called *mercuric oxide cells*, have advantages similar to silver-oxide cells. They're manufactured in the same button-like shape. The main difference, often not of significance, is a somewhat lower voltage per cell: approximately 1.35 V. Since the 1980s, mercury cells and batteries have diminished in popularity because mercury is a toxic heavy metal. When you discard a mercury cell or battery, you must observe special precautions. In some locations, strict laws govern the disposal process, so beware!

TIP   *If you have cells or batteries that you suspect contain mercury, call your local trash-removal department, and get instructions on how to dispose of the components. Don't simply throw them into a rubbish bin!*

Lithium

Lithium cells supply 1.5 V to 3.5 V, depending on the particular processes used in their manufacture. These cells, like their silver-oxide cousins, can be stacked to make batteries. Lithium cells and batteries have superior shelf life, and they can last for years in very-low-current applications. They have exceptional energy capacity per unit of volume. Some engineers believe that lithium batteries will play a key role in the future of electric motorized personal transportation (electric bicycles, motorcycles, cars, boats, and even snow machines).

Lead-Acid

As we've already learned, a lead-acid cell contains a liquid electrolyte of sulfuric acid, along with a lead electrode (negative) and a lead-dioxide electrode (positive). We can connect several such cells in series to get a battery capable of providing useful power for several hours. Some lead-acid batteries contain an electrolyte that has been thickened into a paste to reduce the danger of leakage or spillage. These components work well in consumer devices that require moderate current, such as notebook computers, handheld computers, and portable two-way radios. They are also used in *uninterruptible power supplies* (UPSs) that can provide short-term emergency backup power for desktop computer systems.

Nickel

Nickel-based cells and batteries are available in various configurations. *Cylindrical cells* look like size AAA, AA, C, or D zinc-carbon or alkaline cells. Nickel-based button cells find application in cameras, watches, microcomputer-memory backup circuits, and other places where miniaturization is important. *Flooded cells* are used in heavy-duty situations, and can have storage capacity ratings of up to about 1000 Ah. They have a box-like or carton-like appearance. *Spacecraft cells* are manufactured in strong, thermally insulated packages that can withstand extraterrestrial temperatures and pressures.

Nickel-cadmium (NICAD) batteries are available in box-shaped packages that can be plugged into equipment to form part of the case for a device. An example is the battery pack for a handheld communications transceiver for amateur, citizens-band, police, or military use. This type of battery should never be left connected to a load after it has discharged. If you make that mistake, the polarity of one or more cells might reverse. Once that happens, the battery will no longer accept a recharge, and you must discard it.

Nickel-metal-hydride (NiMH) cells and batteries can directly replace NICAD units in most applications. Environmentally conscious consumers prefer NiMH components over NICADs because the NiMH chemistry does not contain cadmium, which, like mercury, acts as a toxin. Some engineers believe that NiMH batteries also exhibit better behavior than NICADs do when repeatedly and frequently discharged and recharged.

Nickel-based cells and batteries, particularly the NICAD type, sometimes exhibit a bothersome characteristic called *memory* or *memory drain*. If you use such a device repeatedly, and if you allow it to discharge to roughly the same extent with every cycle, it seems to lose most of its energy-storage capacity and "die too soon." You can sometimes "cure" a nickel-based cell or battery of memory drain by discharging it almost completely, recharging it, discharging it almost completely again, and repeating the cycle numerous times. In stubborn cases, however, you'll probably want to buy a new cell or battery instead of trying to rejuvenate the old one.

Nickel-based cells and batteries work best if used with charging units that take several hours to fully replenish the charge. So-called *quick chargers* are available, and their manufacturers might make extraordinary claims as to how fast they can charge up a cell or battery. However, some quick chargers can force too much current through a cell or battery, causing permanent loss of energy-storage capacity. You'll always get the best results if the charger has been made especially for the cell or battery type in question.

TIP   *When a NICAD or NiMH cell or battery has discharged almost all the way, you should fully recharge it as soon as possible. Otherwise, the component might permanently lose most or all of its energy-storage capacity as a result of cell-polarity reversal.*

PROBLEM 5-4

Suppose that you want to make a 9-V battery by stacking silver-oxide cells. How many cells will it take? How can you be sure the polarity is correct?

SOLUTION

A single silver-oxide cell produces 1.5 V, so you'll need six of them to make a 9-V battery. Connected in series, the six cells produce 6 × 1.5 V, which equals 9 V. You must stack up the individual cells so that they're all oriented in the same direction (Fig. 5-6). You should also pay attention to the polarity when using the battery. You'll see one face of each button cell labeled with either a plus sign (+) or a minus sign (−).

TIP   *If you find a cell or battery that lacks labeling to indicate which end is positive and/or which end is negative, you should test it for polarity with a* volt-ohm-milliammeter (VOM), *also called a* multimeter. *You can buy a VOM for a few dollars at hardware stores or hobby electronics stores, such as Radio Shack.*

PROBLEM 5-5

Suppose that you can't find silver-oxide button cells, but lithium button cells are available, each of which provides 1.8 V. How many of these will you have to stack up in series to make a 9-V battery?

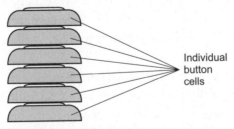

Individual button cells

FIGURE 5-6 · Illustration for Problem 5-4.

✔ SOLUTION

To calculate the exact number of cells, you can divide 9 V by 1.8 V, which represents the voltage provided by a single lithium cell. When you do the arithmetic, you get

$$(9\,V)/(1.8\,V) = 5$$

You'll need five of the button cells in a series "stack" to get the desired voltage in this case.

Photovoltaic Cells and Batteries

A *photovoltaic (PV) cell* is a *semiconductor device* that converts visible light, infrared (IR) radiation, or ultraviolet (UV) radiation directly into electricity. It's also commonly known as a *solar cell*. Engineers combine multiple PV cells to make *photovoltaic batteries* that produce considerable power output in full daylight.

Operation

Figure 5-7 illustrates the internal structure of a silicon PV cell. The device contains two types of specially treated *semiconductor material* called *P type silicon* and *N type silicon*. The top of the assembly and the P type silicon layer are

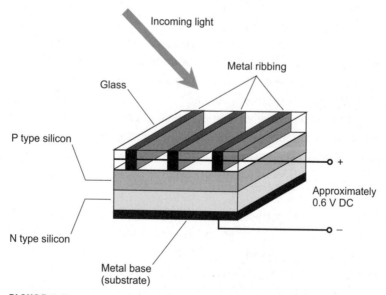

FIGURE 5-7 · Construction of a silicon photovoltaic cell.

transparent, so that IR rays, visible light, or UV rays can pass through easily. Strips of metal *ribbing*, which form the positive electrode, are interconnected with tiny wires. The negative electrode comprises a metal backing called the *substrate*, placed in physical and electrical contact with the N type silicon.

When visible light, IR rays, or UV rays strike the flat boundary between the P type and N type silicon (known as the *P-N junction*), a potential difference develops across the junction. The intensity of the current from the cell, under constant load conditions, varies in direct proportion to the brightness of the incident radiation, up to a certain point. Beyond that point, the current increase becomes more gradual, ultimately leveling off at a maximum value called the *saturation current*. The ratio of the available output power to the incident light power is called the *conversion efficiency*.

Silicon solar cells produce about 0.6 V DC when exposed to light of sufficient brilliance. If the current demand remains low, it doesn't take much light to produce the full output voltage. As the current demand increases, more light is needed to get full output voltage. An upper limit exists to the current that any PV cell can provide, no matter how bright the light gets. This limit depends on the surface area of the cell.

? Still Struggling

If you need to get more current than a single PV cell can deliver, you can connect multiple identical PV cells in parallel to get a *parallel PV battery*. If you need more voltage than a single cell can produce, you can connect multiple identical PV cells in series to get a *series PV battery*. If you need more current *and* more voltage than a single cell can deliver, you can connect multiple identical series combinations of cells in parallel, or multiple identical parallel combinations of cells in series, thereby obtaining a *series-parallel PV battery*.

Solar Panels

Photovoltaic batteries, derived from series-parallel combinations of PV cells, represent the most common construction method for *solar panels*. Combinations that contain exceptionally large numbers of cells are called *PV matrixes* or *PV arrays*. For example, 50 parallel-connected sets of 24 series-connected cells can provide 14 to 15 V (in theory) with substantial deliverable current. Some solar panels have surface areas covering hundreds of square meters.

Although high voltages (such as 500 V) can theoretically result from connecting many photovoltaic cells in series, this "trick" doesn't work very well in practical engineering situations because the internal resistances of the cells add up in series, limiting the current output and causing voltage-regulation problems.

If you want to obtain high voltage from a solar panel, you can use a device called a *power inverter* along with a high-capacity rechargeable electrochemical battery that provides roughly 12 to 48 V. You charge the electrochemical battery with the PV battery, and then use the inverter to convert the low-voltage DC to high-voltage AC. Then, if you want, you can use a rectifier and filter to convert the AC to high-voltage DC.

TIP   *Power inverters are available at large department stores, home supply stores, electronics hobby stores, and hardware stores. Most of them are designed to produce 117 V utility AC from 12-V DC sources such as automotive batteries.*

Solar-Electric Power

You can combine PV batteries and electrochemical storage batteries to build a *stand-alone solar electric energy system*. It uses a large solar panel, a large-capacity lead-acid or nickel-based battery, a power inverter to convert the low-voltage DC into utility AC, and a sophisticated charging circuit for the electrochemical storage battery. Such a system is well suited to environments where the sun shines for a high percentage of the time. This type of system will continue to work even if the utility power grid fails indefinitely.

Solar cells, either alone or supplemented with rechargeable batteries, can be incorporated into an *interactive solar electric energy system*, also called a *grid-intertie system*. This type of system requires a special electrical and financial arrangement with the electric utility company. When the solar power system can't provide for the needs of the household or business, the utility company sells you the deficit. Conversely, when the solar power system produces more than enough energy, the utility company buys the surplus from you.

PROBLEM 5-6

Suppose that you have access to an unlimited supply of identical silicon PV cells. Each cell supplies 0.6 V DC at up to 50 mA in direct sunlight. You want to produce a power source sufficient to operate a device that requires 12 V DC at 1 A. Describe the smallest possible PV array that will accomplish this task, assuming you have an environment where bright, direct sunlight shines when you need to use the device.

✔ SOLUTION

You need a source that will supply 12 V DC, so you must connect sets of 20 PV cells in series. That's true because 20 × 0.6 V = 12 V. When two or more identical PV cells are connected in series, the maximum deliverable current of the combination equals that of any one of the cells, in this case 50 mA. In order to get 1 A of current, you'll have to combine 20 of your series PV sets in parallel. That's true because 20 × 50 mA = 1000 mA = 1 A. The resulting series-parallel array will contain 400 PV cells, forming a so-called 20-by-20 matrix.

▢ PROBLEM 5-7

In the above-described situation, suppose that you build sets of 20 PV cells in parallel, and then combine 20 of these parallel sets in series. Will this arrangement work as well as the system described in the solution to Problem 5-6?

✔ SOLUTION

This scheme will work just as well as the arrangement described in the solution to Problem 5-6. Instead of having 20 sources of 12 V, each capable of delivering 50 mA and all connected in parallel, you'll have 20 sources of 0.6 V, each capable of delivering 1 A and all connected in series. As before, you'll have a 20-by-20 matrix of PV cells. The two PV matrices, while different in terms of schematic geometry, will produce identical results in practice.

Fuel Cells

In the late 1900s, a new type of electrochemical power device emerged that some scientists and engineers believe holds great promise as an alternative energy source: the *fuel cell*.

Hydrogen Fuel

The most talked-about fuel cell during the early years of research and development became known as the *hydrogen fuel cell*. As its name implies, it derives electricity from hydrogen. The hydrogen combines with oxygen (it *oxidizes*) to form energy and water vapor. No pollution occurs, and no toxic by-products come out of the cell. When a hydrogen fuel cell "runs out of juice," you need nothing more than a new supply of hydrogen. The oxygen derives from the earth's atmosphere.

Instead of combusting (burning), the hydrogen in a fuel cell oxidizes in a controlled fashion, and at a much lower temperature than a typical fire. Engineers have devised various oxidation schemes. The *proton exchange membrane (PEM) fuel cell* is one of the most widely used systems. A PEM hydrogen fuel cell generates approximately 0.7 V DC. In order to obtain higher voltages, you can connect fuel cells in series. A series-connected set of fuel cells technically forms a battery, but the more often-used term is *stack*. A stack about the size and weight of a "window type" air conditioner can power a compact electric car. Stacks of small cells called *micro fuel cells* can provide DC to run devices such as portable radios, lanterns, and computers.

Other Fuels

Hydrogen isn't the only chemical that can operate a fuel cell. Almost anything that will combine with oxygen to form energy has been considered as a potential fuel source. *Methanol*, a form of alcohol, has the advantage of being easier to transport and store than hydrogen because it exists as a liquid at room temperature. *Propane* has also been used for powering fuel cells. People commonly store propane as a volatile liquid in tanks for use with barbecue grills and some rural home heating systems. *Methane*, also known as "natural gas," has been used. Even *gasoline* can work!

? Still Struggling

Some scientists and engineers dislike methanol, propane, methane, and gasoline because society has developed dependence on them that "green energy" purists hope to overcome with the implementation of alternative energy systems. In addition, methanol, propane, methane, and gasoline derive from so-called *fossil fuel* sources, the supplies of which, however great they might seem today, are nevertheless finite.

A Promising Technology

As of this writing, fuel cells haven't yet begun to supplant conventional electrochemical cells and batteries in practical electrical applications, largely because of their high cost. Hydrogen is the most abundant and simplest chemical element in the universe, and it produces no toxic by-products. These properties would at first seem to make hydrogen the ideal choice for use in fuel cells. But storage and transport of hydrogen has proven difficult and expensive, especially for fuel cells and stacks intended for operating systems not fixed to permanent pipelines.

An interesting scenario, suggested by one of my physics teachers in the 1970s, involves piping hydrogen gas through standard utility networks designed to carry methane gas. Some modification of the existing system would be required in order to safely handle hydrogen, which is far less dense than methane, and which escapes through small openings easily. But piped-in hydrogen, if obtained at reasonable cost and in abundance, could operate fuel-cell stacks for households and businesses. The DC from such a stack could be converted to utility AC with power inverters. The entire system would occupy roughly the same volume, and would have about the same mass, as a conventional methane-based or propane-based furnace.

QUIZ

This is an "open book" quiz. You may refer to the text in this chapter. You'll find the correct answers listed in the back of the book.

1. If we connect five identical flashlight cells in parallel with all the polarities in agreement, we should expect the combination to produce approximately
   A. 1.5 V DC.
   B. 7.5 V DC.
   C. 0.3 V DC.
   D. 0.0 V DC.

2. Imagine that we charge up a secondary cell to its full extent, and then connect it to a load that draws a continuous current of 1.6 A. This amount of current flows steadily through the load for exactly four hours, and then the cell suddenly "dies." The storage capacity of this cell was evidently
   A. 0.4 Ah.
   B. 1.6 Ah.
   C. 3.2 Ah.
   D. 6.4 Ah.

3. An ideal cell exhibits
   A. a potential difference of 1.018 V.
   B. a flat discharge curve.
   C. high internal resistance.
   D. large storage capacity.

4. A disadvantage of hydrogen as a fuel source is the fact that it
   A. produces water vapor when it burns.
   B. produces toxic compounds when it oxidizes.
   C. is difficult to store.
   D. All of the above

5. Which of the following factors represents a major disadvantage of mercury cells and batteries?
   A. Chemical toxicity
   B. Memory drain
   C. Unstable voltage
   D. Short shelf life

6. When exposed to bright sunlight, a battery comprising three silicon PV cells in series should produce approximately
   A. 0.6 V DC.
   B. 0.2 V DC.
   C. 1.8 V DC.
   D. 5.4 V DC.

7. In which of the following devices does battery polarity make no difference?
    A. A transistor radio
    B. A personal computer
    C. A cell phone
    D. A flashlight

8. Which of the following statements concerning a grid-intertie PV electric energy system holds true?
    A. Power inverters can convert the DC to useful AC.
    B. When the solar power system can't provide enough energy, the utility company sells the deficit.
    C. When the solar power system produces more than enough energy, the utility company buys the surplus.
    D. All of the above

9. In a lead-acid electrochemical battery, the acid constitutes the
    A. load.
    B. electrolyte.
    C. anode.
    D. cathode.

10. We can define an alkaline electrochemical cell's maximum deliverable current as the highest current that it can drive through a load without a significant
    A. reduction in the output voltage.
    B. drop in the internal resistance.
    C. problem with memory drain.
    D. polarity reversal.

Test: Part I

Do not refer to the text when taking this test. You may draw diagrams or use a calculator if necessary. A good score is at least 38 correct. You'll find the answers listed in the back of the book. Have a friend check your score the first time, so you won't memorize the answers if you want to take the test again.

1. Imagine that we have an unlimited supply of new "flashlight" cells. Each cell produces 1.5 V DC, and can deliver up to 7.0 A of current. If we place five of these cells in series with the polarities all in agreement (that is, with positive terminals connected to negative terminals all along the sequence), we'll get a battery that supplies

    A. 1.5 V DC, and can deliver up to 7.0 A of current.
    B. 7.5 V DC, and can deliver up to 7.0 A of current.
    C. 1.5 V DC, and can deliver up to 35 A of current.
    D. 7.5 V DC, and can deliver up to 35 A of current.
    E. None of the above

2. Suppose that we rearrange the cells in the situation of Question 1 so that they're all connected in parallel instead of in series. If we make certain that the polarities all agree (that is, all the positive terminals connect together and all the negative terminals connect together), we'll get a battery that supplies

    A. 1.5 V DC, and can deliver up to 7.0 A of current.
    B. 7.5 V DC, and can deliver up to 7.0 A of current.
    C. 1.5 V DC, and can deliver up to 35 A of current.
    D. 7.5 V DC, and can deliver up to 35 A of current.
    E. None of the above

3. Figure Test I-1 shows five resistors R_1 through R_5, all connected to a battery. The circuit also contains a voltmeter V and an ammeter A. As shown here, the voltages across the individual resistors

    A. all equal the battery voltage.
    B. all equal 1/5 of the battery voltage.
    C. all equal five times the battery voltage.
    D. are directly proportional to the individual resistance values.
    E. are inversely proportional to the individual resistance values.

4. As shown by Fig. Test I-1, meter A is connected specifically to tell us the electricity that "flows"

    A. out of the battery, but not through any of the resistors.
    B. through the combination of R_1 and R_2, but not through R_3, and R_4, or R_5.
    C. through the combination of R_3, R_4, and R_5, but not through R_1 or R_2.
    D. through the combination of all five resistors.
    E. through nothing at all; it will always indicate zero.

5. As shown by Fig. Test I-1, meter V is connected specifically to tell us the electricity that exists

    A. across R_1 and R_2, but not across R_3, R_4, or R_5.
    B. across the combination of R_3, R_4, and R_5, but not across R_1 or R_2.
    C. across any or all of the resistors R_1 through R_5.
    D. across the battery, but not across any of the resistors.
    E. across nothing at all; it will always indicate zero.

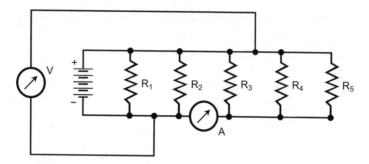

FIGURE TEST I-1 • Illustration for Part I Test Questions 3 through 5.

6. **Figure Test I-2 shows a battery, a switch, and a bulb connected together in**
   A. series.
   B. parallel.
   C. series-parallel.
   D. cascade.
   E. opposition.

7. **Suppose that the battery in the circuit of Fig. Test I-2 provides 13.5 V DC, while the bulb is designed to shine at full brilliance when provided with 13.5 V DC. As the circuit appears here, we should expect**
   A. the bulb to remain dark, and the battery to deliver no current.
   B. the bulb to remain dark, and the battery to deliver some current.
   C. the bulb to glow at partial brilliance, and the battery to deliver no current.
   D. the bulb to glow at full brilliance, and the battery to deliver no current.
   E. the bulb to glow at full brilliance, and the battery to deliver some current.

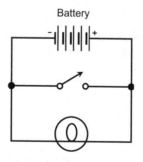

Battery

FIGURE TEST I-2 • Illustration for Part I Test Questions 6 through 9.

8. **If we close the switch in the system shown by Fig. Test I-2, we should expect**

   A. the battery to deliver a lot of current, and the bulb to remain dark.
   B. the battery to deliver no current, and the bulb to remain dark.
   C. the battery to deliver no current, and the bulb to glow at partial brilliance.
   D. the battery to deliver a lot of current, and the bulb to glow at full brilliance.
   E. the battery to deliver a lot of current, and the bulb to glow at partial brilliance.

9. **What, if anything, presents a potential problem or hazard with the system shown in Fig. Test I-2?**

   A. No problem or hazard exists.
   B. If we leave the switch open, the battery could leak, rupture, or explode, but no danger exists for the bulb.
   C. If we close the switch, the battery could leak, rupture, or explode, but no danger exists for the bulb.
   D. If we leave the switch open, the bulb could burn out, but no danger exists to the battery.
   E. If we leave the switch open, the bulb could burn out, and the battery could leak, rupture, or explode.

10. **Which of the following particles never has any electrical charge?**

   A. Atom
   B. Electron
   C. Proton
   D. Neutron
   E. Shell

11. **If two objects both have electron surpluses and we bring them close to each other, we should expect that they will**

   A. lose their charge immediately.
   B. experience a mutual force of repulsion.
   C. experience a mutual force of attraction.
   D. attain equal amounts of charge.
   E. experience no mutual force.

12. **If we connect five 2.0-ohm resistors in parallel, then the *conductance* (not the resistance) of the combination equals**

   A. 2.5 S.
   B. 5.0 S.
   C. 10 S.
   D. 4.0 S.
   E. 0.4 S.

13. **Figure Test I-3 shows two atoms in a good electrical conductor, such as copper wire. A particle moves from Atom 1 to Atom 2, thereby causing an electrical current. The particle that moves in this scenario can be**

   A. a proton.
   B. a neutron.

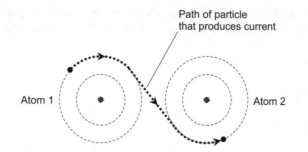

Path of particle
that produces current

Atom 1   Atom 2

FIGURE TEST I-3 · Illustration for Part I Test Questions
13 and 14.

C. an electron.
D. a nucleus.
E. Any of the above

14. **In the situation shown by Fig. Test I-3, the dashed circles represent**

A. proton paths.
B. elementary charge units.
C. electric dipoles.
D. atomic nuclei.
E. shells.

15. **If we connect a 33.0-ohm resistor in series with an 82.0-ohm resistor, the total resistance equals**

A. 23.5 ohms.
B. 49.0 ohms.
C. 52.0 ohms.
D. 57.5 ohms.
E. 115 ohms.

16. **If we connect a 33.0-ohm resistor in parallel with an 82.0-ohm resistor, the total resistance equals**

A. 23.5 ohms.
B. 49.0 ohms.
C. 52.0 ohms.
D. 57.5 ohms.
E. 115 ohms.

17. **An electron is a common example of**

A. a chemical element.
B. a positively charged particle.
C. a neutral atom.
D. an electrical charge carrier.
E. an electric dipole.

18. Figure Test I-4 shows a resistor of ohmic value R connected across a battery that provides a voltage E, causing a current I to flow. If $E = 7.50$ V and $R = 375$ ohms, then

   A. $I = 50.0$ μA.
   B. $I = 2.81$ mA.
   C. $I = 20.0$ mA.
   D. $I = 281$ mA.
   E. $I = 500$ mA.

19. If we increase E by a factor of 9 in the circuit of Fig. Test I-4, while leaving R constant, then I will

   A. increase by a factor of 81.
   B. increase by a factor of 9.
   C. increase by a factor of 3.
   D. decrease by a factor of 3.
   E. decrease by a factor of 81.

20. Suppose that, in the circuit of Fig. Test I-4, $E = 12.0$ V and $I = 1.50$ mA. From this information, we know that

   A. $R = 80.0$ ohms.
   B. $R = 180$ ohms.
   C. $R = 1.25$ k.
   D. $R = 1.80$ k.
   E. $R = 8.00$ k.

21. If we leave E constant in the circuit of Fig. Test I-4 but we want to reduce the value of I to 1/16 of its former value, then we must make R

   A. 1/16 as great as before.
   B. 1/4 as great as before.
   C. four times as great as before.
   D. 16 times as great as before.
   E. 64 times as great as before.

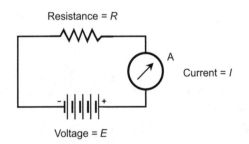

Resistance = R

A

Current = I

Voltage = E

FIGURE TEST I-4 • Illustration for Part I Test Questions 18 through 23.

22. Suppose that, in the circuit of Fig. Test I-4, we set $R = 1.80$ M and then we observe that $I = 600$ nA. From this information, we know that

    A. $E = 1.08$ V.
    B. $E = 333$ mV.
    C. $E = 30.0$ mV.
    D. $E = 10.8$ mV.
    E. $E = 3.33$ mV.

23. If we want to maintain a constant current I in the circuit of Fig. Test I-4 as we reduce R to 1/10 of its previous value, then we must make E

    A. 100 times as great as before.
    B. 10 times as great as before.
    C. 1/10 as great as before.
    D. 1/100 as great as before.
    E. 1/1000 as great as before.

24. The maximum amount of current that we can get from a lead-acid cell depends primarily on the cell's

    A. mass and volume.
    B. voltage.
    C. polarity.
    D. shelf life.
    E. All of the above

25. The energy storage capacity of an electrochemical cell or battery in watt-hours equals the ampere-hour capacity multiplied by the

    A. maximum deliverable current.
    B. cell or battery voltage.
    C. length of time that current flows.
    D. output power in watts.
    E. load resistance in ohms.

26. If an electrochemical battery produces output current that gradually goes down when we connect a constant load resistance to it, we can say that the battery exhibits

    A. a flat discharge curve.
    B. a linear discharge curve.
    C. a proportional discharge curve.
    D. an inverse discharge curve.
    E. a declining discharge curve.

27. If two objects have opposite electric charges (one negative, the other positive) and we bring them close to each other, we should expect that they will

    A. lose their charge immediately.
    B. experience a mutual force of repulsion.
    C. experience a mutual force of attraction.
    D. attain equal amounts of charge.
    E. experience no mutual force.

28. Suppose that we connect a 27.00-V battery as the DC voltage source in the system of Fig. Test I-5. We close the switch and read the ammeter, which tells us that I = 54.00 mA. Based on this information, we know that the potentiometer dissipates

    A. 50.00 μW.

    B. 2.000 mW.

    C. 500.0 mW.

    D. 1.458 W.

    E. 39.37 W.

29. Suppose that we leave the switch closed in the system of Fig. Test I-5 and Question 28. Then we increase the DC source voltage by a factor of 25, and simultaneously increase the potentiometer's resistance by a factor of 25. The power that the potentiometer dissipates will

    A. increase by a factor of 5.

    B. increase by a factor of 25.

    C. increase by a factor of 125.

    D. decrease by a factor of 5.

    E. decrease by a factor of 25.

30. Suppose that we connect a 4.500-V battery as the DC voltage source in the system of Fig. Test I-5. We close the switch and read the ammeter, which tells us that I = 90.00 mA. We leave the switch closed for precisely 4 hours, timed right down to the second. Based on this information, we know that during this period of time, the potentiometer dissipates

    A. 1.620 Wh.

    B. 4.050 Wh.

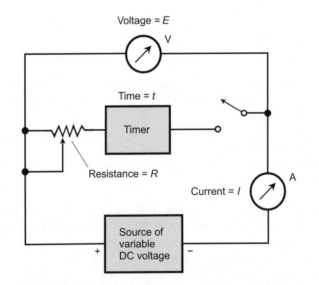

FIGURE TEST I-5 · Illustration for Part I Test Questions 28 through 31.

C. 32.40 Wh.

D. 64.80 Wh.

E. 162.0 Wh.

31. Suppose that we repeat the experiment described in Question 30, but with an 18.0-V battery instead of a 4.50-V battery. We don't change the potentiometer's resistance. In precisely 4 hours, timed right down to the second, the potentiometer will dissipate

A. 3.240 Wh.

B. 6.480 Wh.

C. 25.92 Wh.

D. 2.291 Wh.

E. 259.2 mWh.

32. Imagine two charged spherical objects. Assume that the charge is distributed uniformly throughout each sphere. The left-hand sphere contains 4 C of positive charge, and the right-hand sphere contains 8 C of negative charge. We observe an electrostatic force F between the two spheres. Now suppose that we double the charge quantity on the left-hand sphere, making it 8 C, but we do nothing to the charge on the right-hand sphere. If we keep the distance between the two spheres constant, the electrostatic force between the spheres will

A. stay the same.

B. increase to 2F.

C. increase to 4F.

D. decrease to F/2.

E. decrease to F/4.

33. Figure Test I-6 shows four batteries connected in series. Pay careful attention to the polarity of each battery! In this situation, we can expect that

A. $E = 12$ V.

B. $E = 9.0$ V.

C. $E = 7.5$ V.

D. $E = 6.0$ V.

E. $E = 4.5$ V.

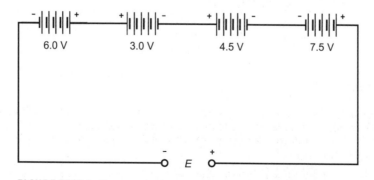

FIGURE TEST I-6 · Illustration for Part I Test Questions 33 and 34.

34. If we rearrange the batteries in the scenario of Fig. Test I-6 so that the polarities all agree and each negative battery terminal faces to the left, we can expect that

    A. $E = 21$ V.
    B. $E = 18$ V.
    C. $E = 15$ V.
    D. $E = 12$ V.
    E. $E = 10$ V.

35. When a nickel-based battery has discharged almost all the way, we should fully recharge it as soon as possible to avoid

    A. overheating and possible fire risk.
    B. chemical leakage from the battery housing.
    C. loss of internal resistance.
    D. loss of energy-storage capacity as a result of cell-polarity reversal.
    E. All of the above

36. Which of the following combustible substances cannot, in theory, power a fuel cell?

    A. Methanol
    B. Hydrogen
    C. Propane
    D. Gasoline
    E. Any of the above substances can, in theory, power a fuel cell.

37. Figure Test I-7 illustrates a network of resistors connected to a common central branch point. The arrows denote the directions of conventional current flow. If all three of the upper resistors carry the same current, then we can correctly replace each of the three query symbols (?) with the notation

    A. 2.50 A.
    B. 4.00 A.
    C. 5.00 A.
    D. 6.00 A.
    E. 15.0 A.

38. Figure Test I-8 shows a network of four resistors connected in series with a battery. How much potential difference appears across the resistor marked with the query symbol (upper right)?

    A. 12.5 V
    B. 13.5 V
    C. 14.5 V
    D. 21.0 V
    E. 25.0 V

39. Suppose that we double the battery voltage in the circuit of Fig. Test I-8, from 45.0 V to 90.0 V, but we leave all four resistance values the same. In this case how much potential difference will appear across the resistor marked with the query symbol (upper right)?

    A. 12.5 V
    B. 13.5 V

All three unknown current values are equal

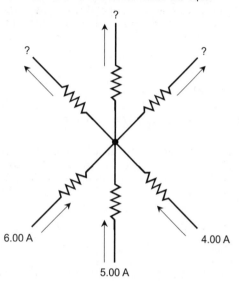

FIGURE TEST I-7 • Illustration for Part I Test
Question 37.

C. 14.5 V
D. 21.0 V
E. 25.0 V

40. **Suppose that we connect eight identical light bulbs in series with a large battery. All of the bulbs light up to half brilliance. If we remove one of the bulbs and don't put anything in its place, what will happen to the amount of electricity that each of the other seven bulbs receives?**

A. It will decrease slightly.
B. It will remain the same.
C. It will increase slightly.
D. It will drop to zero.
E. We need more information to answer this question.

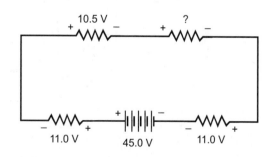

FIGURE TEST I-8 • Illustration for Part I Test
Questions 38 and 39.

41. Suppose that we connect eight identical light bulbs in series with a large battery. All of the bulbs light up to half brilliance. If we add another bulb in series with all the rest (and identical to all the rest), what will happen to the amount of electricity that each of the original eight bulbs receives?

  A. It will decrease slightly.
  B. It will remain the same.
  C. It will increase slightly.
  D. It will drop to zero.
  E. We need more information to answer this question.

42. Suppose that we connect eight identical light bulbs in series with a large battery. All of the bulbs light up to half brilliance. If we place a short circuit across one of the bulbs, what will happen to the amount of electricity that each of the other seven bulbs receives?

  A. It will decrease slightly.
  B. It will remain the same.
  C. It will increase slightly.
  D. It will drop to zero.
  E. We need more information to answer this question.

43. Figure Test I-9 illustrates

  A. an alkaline cell.
  B. a cadmium cell.

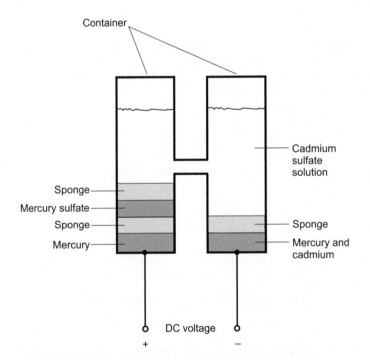

FIGURE TEST I-9 · Illustration for Part I Test Questions 43 and 44.

C. a mercury cell.

D. a standard cell.

E. a sulfate cell.

44. **What's the DC voltage produced by the device shown in Fig. Test I-9?**

A. 8.765 V

B. 6.282 V

C. 3.142 V

D. 1.018 V

E. It depends on the volume of the cell.

45. **Imagine an electrical device (let's call it component X) connected directly to a battery. Suppose that component X has a certain resistance. The voltage from the battery causes some current to flow through X. If we multiply the battery voltage by a factor of 4, and we also multiply the resistance of component X by a factor of 4, the current through X will**

A. stay the same.

B. increase by a factor of 2.

C. increase by a factor of 4.

D. increase by a factor of 16.

E. decrease by a factor of 4.

46. **The standard unit of electric charge quantity is the**

A. ampere.

B. volt.

C. watt.

D. coulomb.

E. joule.

47. **When exposed to bright sunlight, a battery comprising four 0.6-V photovoltaic (PV) cells in parallel should produce approximately**

A. 0.6 V DC.

B. 1.2 V DC.

C. 2.4 V DC.

D. 4.8 V DC.

E. 9.6 V DC.

48. **Figure Test I-10 shows the current that a hypothetical electrochemical cell delivers, plotted as a graph versus time. The area under the curve (that is, the area of the region bounded by the curve and the two coordinate axes) represents the battery's**

A. maximum deliverable current.

B. shelf life.

C. storage capacity.

D. internal resistance.

E. volt-ampere output.

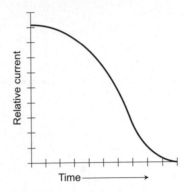

FIGURE TEST I-10 · Illustration for
Part I Test Question 48.

49. **Which of the following statements concerning a stand-alone PV energy system holds true?**
    A. No storage battery is required; the energy can be derived directly from the photovoltaic cells at all times.
    B. When the solar power system can't provide enough energy, the utility company sells the deficit.
    C. When the solar power system produces more than enough energy, the utility company buys the surplus.
    D. The system will work even if the utility grid fails.
    E. All of the above

50. **An elementary charge unit (ECU) equals the electrical charge quantity on**
    A. a single electron.
    B. a single proton.
    C. a single neutron.
    D. a neutral atom.
    E. More than one of the above

Part II

Alternating Current

chapter 6

What Is Alternating Current?

In *alternating current* (AC), the charge carriers (usually electrons) reverse direction at regular intervals. Between the output terminals of an *AC voltage* source, the polarity reverses periodically. Alternating-current appliances consume *AC power* that, over time, accumulates as *AC energy*.

CHAPTER OBJECTIVES

In this chapter, you will

- Define and quantify AC amplitude, frequency, and period.
- Compare sine, square, sawtooth, and complex AC waveforms.
- Learn how to express AC amplitude in various ways.
- See how utility companies generate electricity for public use.
- Learn why AC prevails over DC in electrical transmission systems.

How a Wave Alternates

When we portray an AC wave as a graph of current or voltage versus time, we define a *cycle* as a portion of the wave that lies between a specific point on the graph and the corresponding point on the next alternation. Figures 6-1 A and B illustrate some points that engineers use to mark the instant where a single cycle starts and ends.

Crests and Troughs

Figure 6-1A illustrates two successive *wave crests*. Crests represent points at which the wave attains its maximum positive amplitude. The time required for a single cycle to complete itself corresponds to the distance in the graph between any two adjacent crests. Figure 6-1A also shows successive *wave troughs*, points at which the wave attains its maximum amplitude opposite the direction defined as positive. As with crests, the time period for one cycle corresponds to the distance between two successive troughs.

Zero Points

We can define, express, or measure one cycle of a wave by defining, expressing, or measuring the time interval between any two successive points where the

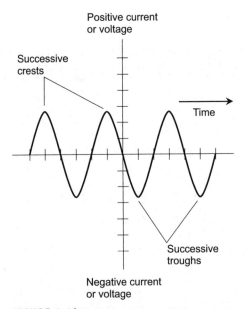

FIGURE 6-1A · Successive crests and troughs in a wave can define a cycle.

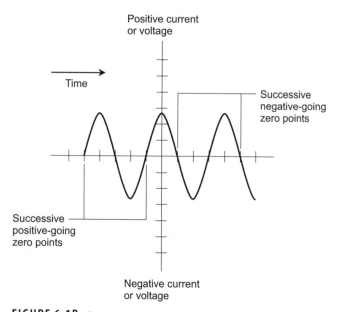

Positive current or voltage

Time

Successive negative-going zero points

Successive positive-going zero points

Negative current or voltage

FIGURE 6-1B · Successive zero points in a wave can define a cycle.

wave crosses the time axis. Because any point on the time axis indicates an amplitude of zero, we can call it a *zero point*. The zero points correspond to a momentary absence of any current, voltage, or power. Figure 6-1B shows two successive *positive-going zero points* and two successive *negative-going zero points*.

TIP *The AC waves in Figs. 6-1 A and B are* **sine waves,** *also called* **sinusoids.** *This name derives from the fact that such a wave shows up as some variant of the* **trigonometric** *sine function when we graph its amplitude as a function of time. A sinusoid represents the most perfect possible AC wave, in the sense that it concentrates all of its energy at a single frequency. No other wave type has this property.*

Instantaneous Amplitude

The *instantaneous amplitude* of an AC wave is the current, voltage, or power at some precise moment in time. The instantaneous amplitude of the AC from a utility outlet constantly and rapidly changes, in contrast to the instantaneous amplitude of the electricity from a battery, which remains constant as time passes (except, in some cases, for a gradual decline as the battery discharges).

The way in which the instantaneous amplitude of an AC wave varies depends on the shape of the graph of the wave. This *wave shape* is also called the *waveform*. On a graph, we can show instantaneous amplitudes for a particular AC wave as individual points at various instants in time. That's how we get the term "instantaneous."

Period

The time interval between any two successive wave crests, troughs, positive-going zero points, or negative-going zero points is called the *period* of the AC wave. In equations and formulas, we symbolize the period as an uppercase italic letter T. The standard unit for the period is the *second*, which engineers and scientists abbreviate as a lowercase, nonitalic letter s.

In theory, an AC wave can have a period as short as a tiny fraction of a second, or as long as billions of years. In most electrical circuits, T equals a small fraction of 1 s. For example, the period of a typical household utility AC wave in the United States equals 1/60 s.

Period is often expressed in power-of-10 parts a second such as *milliseconds* (ms), *microseconds*, (μs), *nanoseconds* (ns), and *picoseconds* (ps). We can define these time units as follows:

$$1 \text{ ms} = 0.001 \text{ s}$$
$$= 10^{-3} \text{ s}$$
$$1 \text{ μs} = 0.000001 \text{ s}$$
$$= 10^{-6} \text{ s}$$
$$1 \text{ ns} = 0.000000001 \text{ s}$$
$$= 10^{-9} \text{ s}$$
$$1 \text{ ps} = 0.000000000001 \text{ s}$$
$$= 10^{-12} \text{ s}$$

Frequency

The *frequency* of an AC wave, denoted by the lowercase italic letter f, equals the mathematical reciprocal of the period. We can express this fact as the equation

$$f = 1/T$$

or as

$$T = 1/f$$

Prior to roughly the year 1970, technical people expressed frequency in units called *cycles per second*, abbreviated cps. High frequencies were expressed in *kilocycles* (kc), *megacycles* (Mc), *gigacycles* (Gc) or *teracycles* (Tc), representing thousands, millions, billions (thousand-millions), or trillions (million-millions) of cycles per second.

Nowadays, the standard unit of frequency is the *hertz*, abbreviated Hz. The hertz represents the same parameter as the cycle per second does. For example, 1 Hz = 1 cps, 10 Hz = 10 cps, and 57,338,271 Hz = 57,338,271 cps. In most "real-world" scenarios, high AC frequencies are expressed in *kilohertz* (kHz), *megahertz* (MHz), *gigahertz* (GHz), or *terahertz* (THz). The relationships proceed as follows:

$$1 \text{ kHz} = 1000 \text{ Hz}$$

$$= 10^3 \text{ Hz}$$

$$1 \text{ MHz} = 1,000,000 \text{ Hz}$$

$$= 10^6 \text{ Hz}$$

$$1 \text{ GHz} = 1,000,000,000 \text{ Hz}$$

$$= 10^9 \text{ Hz}$$

$$1 \text{ THz} = 1,000,000,000,000 \text{ Hz}$$

$$= 10^{12} \text{ Hz}$$

PROBLEM 6-1

Suppose that your friend says her old computer has a microprocessor clock speed of 500 MHz. What's this frequency in hertz? Express the answer as a number written out in full, and also as a number in power-of-10 notation.

✔ SOLUTION

Note that 1 MHz = 10^6 Hz, or 1,000,000 Hz. Therefore, 500 MHz equals 500 times this amount. You calculate

$$500 \text{ MHz} = (500 \times 1,000,000) \text{ Hz}$$

$$= 500,000,000 \text{ Hz}$$

$$= 5.00 \times 10^8 \text{ Hz}$$

PROBLEM 6-2

What's the period *T* of the wave representing the microprocessor clock signal in the situation of Problem 6-1? Express the answer in seconds, and also in nanoseconds.

☑ **SOLUTION**

In order to find the period in seconds, you must divide 1 by 500,000,000. A calculator can be useful here if it can display enough digits. You calculate

$$T = (1/500,000,000) \text{ s}$$

$$= 0.00000000200 \text{ s}$$

$$= 2.00 \times 10^{-9} \text{ s}$$

$$= 2.00 \text{ ns}$$

TIP   *In both of the preceding solutions, you can justify three significant figures in your final answer because that's the precision of the original frequency value (500 MHz).*

The Shape of a Wave

People usually imagine a smooth, undulating wave when they think of AC, but such a wave can have infinitely many different shapes. Let's look at the most common AC waveforms.

Sine Wave

When displayed on a laboratory oscilloscope or plotted as a graph, a sine wave has a characteristic shape. It always resembles the heavy, undulating curves in Fig. 6-1A or Fig. 6-1B. Any AC wave that concentrates all of the electrical energy at a single, constant, defined frequency is a sinusoid. The converse of this principle also holds true: Any perfect sinusoid concentrates all of the electrical energy at a single, constant, defined frequency.

Square Wave

When portrayed literally as a function of time, a *square wave* looks like a pair of parallel, dashed lines, one with positive polarity and the other with negative polarity. The transitions between negative and positive polarity take place instantaneously, so in theory they shouldn't show up on an oscilloscope (and often do not in actual tests). In engineering graphs, we can draw square-wave transitions as vertical lines, so we get a waveform that looks like the heavy, "meandering" line in Fig. 6-2.

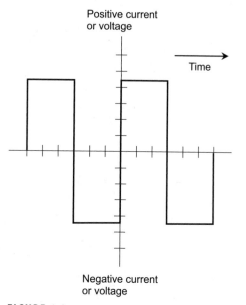

FIGURE 6-2 · A square wave.

Sawtooth Waves

Some AC waves rise and/or fall in straight, but diagonal, lines as seen on an oscilloscope screen. The slope of the line indicates how fast the amplitude changes. We call such AC waveforms *sawtooth waves* because of their appearance. Figure 6-3A shows an example of a sawtooth wave with a finite, measurable *rise time* and a practically instantaneous (zero) *decay time*. Because of its defined, straight-line "up-slope," engineers sometimes refer to this waveform as a *ramp*. Figure 6-3B shows a sawtooth wave with a practically instantaneous rise time and a finite, measurable decay time. It's exactly the reverse situation from the ramp.

Complex Waves

An AC wave can have a complicated shape, sometimes so crazy that it doesn't look like a wave at all. Nevertheless, as long as a varying "signal" has a definite period, and as long as the polarity keeps switching back and forth between positive and negative, we can call it true AC.

Some AC "signals" comprise two or more *component waves* with different frequencies. For example, we might combine three AC sine waves, one with a frequency of 60 Hz, another with a frequency of 120 Hz, and a third with

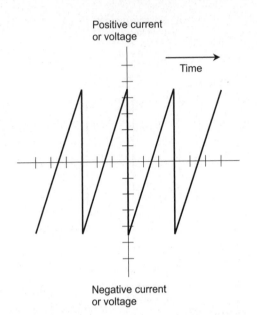

FIGURE 6-3A · A sawtooth wave with a slow rise and a fast decay.

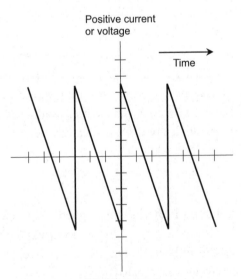

FIGURE 6-3B · A sawtooth wave with a fast rise and a slow decay.

a frequency of 180 Hz, all equally strong. The resultant wave won't look like a sinusoid, but it will have a period.

PROBLEM 6-3

In Fig. 6-3A, suppose that each horizontal division represents 1.00 ns. What's the period of the ramp wave in that case? Assume that the vertical-line portions of the wave (representing the instantaneous negative transitions) pass *precisely* through the vertical divisions as they appear in the drawing.

SOLUTION

The period, T, equals the time between successive crests, troughs, positive-going zero points, or negative-going zero points. We can see that any two successive crests or troughs are three divisions apart. We can also see that the centers of any two consecutive "up-sloping" ramp lines (which pass through the positive-going zero points) are separated by three horizontal divisions, and the vertical negative-going transitions (which pass through the negative-going zero points) are separated by three horizontal divisions. Therefore, $T = 3.00$ ns.

PROBLEM 6-4

What's the frequency of the wave shown in Fig. 6-3A? Express the answer in hertz, gigahertz, and megahertz.

SOLUTION

We've determined that $T = 3.00$ ns, which equals 3.00×10^{-9} s or 0.00000000300 s. Using a calculator, we can determine the approximate frequency f in hertz by finding the reciprocal of T in seconds, as follows:

$$f = 1/T$$
$$= 1/(0.00000000300)$$
$$= 3.33 \times 10^8 \text{ Hz}$$
$$= 0.333 \times 10^9 \text{ Hz}$$
$$= 0.333 \text{ GHz}$$

To find the frequency in megahertz, remember that $1 \text{ MHz} = 10^6 \text{ Hz} = 1,000,000$ Hz. Therefore, the frequency of the wave shown in Fig. 6-3A, assuming that each horizontal division represents 1.00 ns, is approximately 333 MHz.

The Strength of a Wave

Engineers and technicians commonly express the amplitude, also called strength or intensity, of an AC wave in terms of voltage. Sometimes you'll read or hear about wave amplitude in terms of current. Once in awhile, an engineer will talk about a wave's amplitude in terms of the power that it contains or dissipates.

Positive and Negative Peak Amplitude

We define the *positive peak amplitude* of an AC wave as the maximum positive value that the instantaneous amplitude attains. Engineers sometimes abbreviate positive peak amplitude as pk+. We might see the positive peak amplitude of an AC wave expressed as 4.5 V pk+, for example. Similarly, the *negative peak amplitude* (pk−) equals the maximum negative value that the instantaneous amplitude attains.

In many, if not most, AC waves, the positive and negative peak amplitudes have equal and opposite values. But sometimes they differ. Figure 6-4A illustrates the concept of positive and negative peak current or voltage for a pure AC sine wave. In this case, they're equal and opposite.

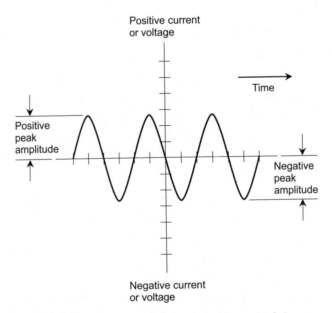

FIGURE 6-4A · Positive peak (pk+) and negative peak (pk−) amplitudes for a pure sine wave.

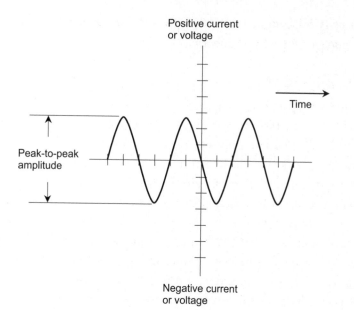

FIGURE 6-4B · Peak-to-peak (pk-pk) amplitude for a pure sine wave.

TIP   *The positive and negative peak amplitudes of an AC wave are independent of the period, and are also independent of the frequency.*

Peak-to-Peak Amplitude

The *peak-to-peak* (pk-pk) *amplitude* of a wave equals the mathematical difference between the positive peak amplitude and the negative peak amplitude, taking polarity into account. We can informally write this fact as

$$\text{pk-pk} = (\text{pk+}) - (\text{pk--})$$

If the positive peak amplitude and the negative peak amplitude have identical extent but opposite polarity, then the peak-to-peak amplitude equals twice the positive peak amplitude or −2 times the negative peak amplitude. That's the case with all pure AC sine waves.

Figure 6-4B shows the notion of peak-to-peak current or voltage for a pure AC sine wave. Peak current has a direction assigned to it, and peak voltage has a polarity. But peak-to-peak quantities lack direction or polarity. Therefore, after calculating the peak-to-peak current or voltage, we remove the directional expression or polarity sign.

TIP   *The peak-to-peak amplitude of an AC wave is independent of the period, and is also independent of the frequency.*

Average Amplitude

We define a wave's *average amplitude* as the set of all possible instantaneous amplitude values, mathematically averaged over exactly one cycle. In a pure AC sine wave with equal but opposite positive and negative peaks, the average amplitude equals zero because the positive and negative half-waves, having identical "mirror-image" shapes, cancel each other out over time. The same phenomenon occurs with a square AC wave, as long as the positive and negative peaks are equal and opposite.

If we have a pure sinusoid but the positive and negative peak amplitudes differ, we can find the average amplitude by calculating the *arithmetic mean* (the mathematical average) of the positive and negative peak values. We add those two values and then divide by precisely 2, remembering to take polarity into account.

An irregular AC wave will sometimes have an average amplitude of zero by sheer happenstance, but more often it will turn out positive or negative. The average amplitude for an irregular wave depends on the shape of the wave, and also on the positive and negative peak amplitudes. An unlimited variety of irregular waveforms can arise in electronic systems, some of them with incredible complexity.

TIP   *The average amplitude of an AC wave, no matter how simple or complicated it might be, is independent of the period, and is also independent of the frequency.*

? Still Struggling

If we superimpose a *DC component* on a perfect AC sine wave, then the average voltage (E_{avg}) of the wave equals the DC component. For example, if we connect a utility AC voltage source in series with a 45-V DC battery, the resulting output exhibits an average voltage of + 45 V avg or −45 V avg, depending on which way we connect the battery. We won't often encounter situations of this sort in household utility circuits, but AC waves with superimposed DC components do occasionally appear in sophisticated electronic devices.

Root-Mean-Square Amplitude

Often, engineers want to express the *effective amplitude* or the *DC-equivalent amplitude* of an AC wave. It's tempting to suppose that a wave's effective or DC-equivalent amplitude equals its average amplitude, but in fact it almost never does! We define the effective amplitude of an AC wave as the voltage, current, or power that a DC source would have to generate to produce the same practical effect as the AC wave in a pure resistance. The most common way to express effective AC amplitude involves the so-called *root-mean-square* (RMS) value.

The mathematical details of RMS determinations get rather arcane. Technically, we can calculate the RMS value of a wave by squaring all of the instantaneous values, then averaging them, and finally taking the square root of the result. We take the *root* of the *mean* of the *squares*; that's where the terminology comes from. In a more "down-to-earth" thought mode, we can see how RMS values compare with other parameters by considering sine waves and square waves specifically.

- For a perfect AC sine wave with no DC component, the RMS current or voltage equals *roughly* 0.3536 times the peak-to-peak current or voltage, and the peak-to-peak current or voltage equals *roughly* 2.828 times the RMS current or voltage.

- For a perfect AC square wave with no DC component, the RMS current or voltage equals *exactly* half the peak-to-peak current or voltage, and the peak-to-peak current or voltage equals *exactly* twice the RMS current or voltage.

For other AC waveforms, the relationship between the RMS value and the peak-to-peak value depends on the shape of the wave.

TIP *The RMS amplitude of an AC wave, like the positive peak, negative peak, peak-to-peak, and average amplitudes, is independent of the period, and is also independent of the frequency.*

TIP *When an electrician says that a wall outlet supplies 117 V, she means to say that it puts out 117 root-mean-square volts (117 V RMS)—not 117 average volts, not 117 positive peak volts, not −117 negative peak volts, and not 117 peak-to-peak volts.*

? Still Struggling

The existence of superimposed DC on a wave affects its RMS amplitude. The task of figuring out the RMS value in such situations can get "mathematically messy" for complicated irregular waveforms, but a good computer program can usually meet the challenge. In any event, the RMS amplitude of a wave, no matter how strange or intricate, can't exceed the positive peak amplitude or −1 times the negative peak amplitude, whichever is larger.

PROBLEM 6-5

Figure 6-5 illustrates a perfect AC sine wave as it might appear on a laboratory oscilloscope. Voltage values appear on the vertical scale, and time appears on the horizontal scale. Each vertical division represents 5.0 V. Time divisions are not quantified. What's the approximate positive peak voltage of this wave? What's its approximate negative peak voltage?

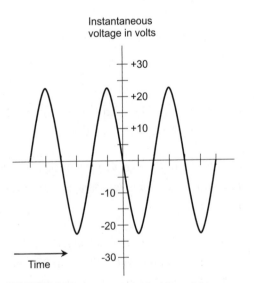

FIGURE 6-5 · Illustration for Problems 6-5 through 6-8.

SOLUTION

A close visual examination of the graph reveals that the positive peak voltage equals about +23 V pk+, and the negative peak voltage equals about −23 V pk−.

PROBLEM 6-6

What's the approximate peak-to-peak voltage of the AC wave shown in Fig. 6-5?

SOLUTION

The peak-to-peak voltage (let's call it E_{pk-pk}) equals the mathematical difference between the positive and negative peak voltages. We can calculate it as follows:

$$E_{pk-pk} = +23 - (-23)$$
$$= +23 + 23$$
$$= 46 \text{ V pk-pk}$$

We don't use any polarity sign in our answer here because it represents a peak-to-peak value for which polarity has no meaning.

PROBLEM 6-7

What's the average voltage of the AC wave shown in Fig. 6-5?

SOLUTION

The average voltage equals 0.0 V, assuming that the positive and negative peak voltages are equal and opposite, and assuming (as we're told in Problem 6-5) that the wave is a perfect sinusoid.

PROBLEM 6-8

What's the RMS voltage of the AC wave shown in Fig. 6-5?

✔ **SOLUTION**

For a sine wave, the RMS value equals approximately 0.3536 times the peak-to-peak value, as long as no DC component exists. In this situation, we have no DC component, so we can calculate the RMS voltage E_{RMS} to two significant figures as

$$E_{RMS} = 0.3536 \times 46$$
$$= 16 \text{ V RMS}$$

How to Produce AC

Engineers can obtain AC electricity by several means, the most common of which is the simple *AC generator*. Another scheme, more often used in residential and small-business scenarios, involves converting DC to AC using a *power inverter*.

The AC Generator

An AC generator produces electricity from a coil that rotates with respect to a powerful magnetic field. Whenever an electrical conductor moves in a magnetic field, electric current arises in that conductor. A generator can comprise a rotating magnet inside a fixed coil of wire, or a rotating coil inside a fixed magnet or set of magnets. A mechanical force drives the rotating magnet or coil shaft. Such force can originate from falling or flowing water, prevailing winds, ocean currents, nuclear reactions, heat from inside the earth, steam under pressure, or explosive combustion of fossil fuels. This type of generator always produces AC directly.

Small portable gasoline-powered generators, capable of outputting a few kilowatts of useful power, can be purchased in department stores. Medium-sized generators can produce enough electricity to supply a small residence or business. The most massive electric generators, found in power plants, are as large as a typical residential house, can produce several megawatts of power continuously, and can provide sufficient electricity for a community.

Whenever we generate AC by rotating a magnet in a coil of wire, or by rotating a coil of wire inside a set of magnets, AC voltage appears between the opposite ends of the wire coil. The voltage depends on the strength of the magnet(s), the number of turns in the wire coil, and the angular (rotational) speed of the magnet(s) or coil. The AC frequency depends only on the angular speed.

Normally, for utility AC in the United States, the angular speed equals a constant, precise 3600 revolutions per minute (RPM) or 60 revolutions per second (RPS), resulting in an AC output frequency of 60 Hz.

TIP   *In some countries, the standard generator rotation speed equals 2500 RPM or 50 RPS, producing an AC output frequency of 50 Hz.*

Generator Efficiency

It doesn't take much rotational force, or *torque*, to turn the shaft of a generator under *no-load conditions* (when you connect nothing at all to its output, so the generator doesn't have to deliver any current). But as soon as you connect a *load*—something that draws current, such as a light bulb or an electric heater—to the output terminals of an AC generator, you'll find it more difficult to turn the shaft.

As users demand more and more electrical power from a generator, it takes more and more mechanical torque to drive it. This fact explains why you can't connect a generator to a stationary bicycle and expect to pedal your whole town into electrification. The electrical power that comes out of a generator can never exceed the mechanical power that drives it. In fact, some energy always gets lost in the electricity-generation process, mainly as heat in the coil, magnets, and external apparatus.

Imagine that you connect a generator to a stationary bicycle. Your legs might provide enough power to run a small radio, but nowhere near enough to supply your entire house. Even with a light load connected to a bicycle-driven generator, you'll find your body getting hot after awhile, as a consequence of the fact that you're "burning" food-energy calories! The generator will get hot as well. All that heat represents power that can't appear as electricity at the generator output.

We define *generator efficiency* as the ratio of the useful electrical power output to the mechanical power input, both measured in the same units (such as watts or kilowatts). We can multiply this ratio by 100 to get a percentage. If P_D represents the mechanical driving power that drives a generator and P_E represents the useful electrical output power in the same units, then the efficiency (*Eff*), as a ratio, can be calculated using the formula

$$Eff = P_E/P_D$$

The efficiency (*Eff$_\%$*), as a percentage, is

$$Eff_\% = 100 \ (P_E/P_D)\%$$

TIP *No matter how expertly engineers design, build, and operate an electric generator, it will inevitably exhibit less than 100% efficiency in the "real world"— sometimes a lot less.*

The Power Inverter

A *power inverter*, sometimes called a *chopper power supply*, delivers high-voltage AC from a low-voltage DC source. The input is typically 12 to 14 V DC, and the output is usually 110 to 130 V RMS AC. In the United States, you can buy a small power inverter in most large department stores or hardware stores. It can operate small devices, such as lamps and low-power hi-fi radios, but it isn't designed to deliver power to large appliances, households, or businesses.

Figure 6-6 shows a simplified block diagram of a power inverter. The *chopper* opens and closes a *switching transistor* at a rate of approximately 60 Hz, interrupting the battery current to produce pulsating DC. The *transformer* converts the pulsating DC to AC, and also steps up the voltage. If the battery is rechargeable, solar panels or a residential wind turbine can replenish its charge and provide a long-term source of utility power.

Low-cost power inverters produce waveforms that don't closely resemble a sine wave because the chopper interrupts the DC source to create a series of rectangular pulses. Rectangular-wave choppers are inexpensive and easy to manufacture. For this reason, low-cost power inverters don't always work well with appliances, such as personal computers and microcomputer-controlled appliances that need a nearly perfect, 60-Hz sine-wave source of AC power.

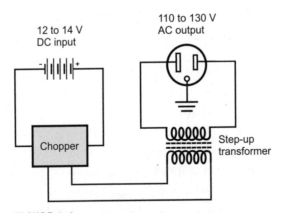

FIGURE 6-6 · A power inverter for converting low-voltage DC to utility AC.

TIP   *Sophisticated (and expensive) power inverters produce fairly good sine waves, and have a frequency close to 60 Hz. This type of inverter represents a good investment if you need to use it to operate sensitive electronic systems.*

Edison versus AC

According to "popular wisdom," the famous American inventor *Thomas Edison* (1847–1931) favored DC over AC for electrical power transmission in the early 1900s, when engineers originally planned and developed the utility power grid. Edison's colleagues argued that AC would work better than DC, and they prevailed. All high-voltage utility lines that you see today carry AC, not DC. Nevertheless, DC offers at least one major advantage over AC in electric power transmission, an asset that becomes apparent when we try to send useful electricity over great distances by wire. Direct current, at an extremely high voltage, propagates along wires more efficiently than AC current does at a similar voltage. The wire offers less effective resistance to the DC than it presents to the AC. As a result, less energy goes to waste as a result of the magnetic fields that always surround current-carrying wires.

?

Still Struggling

Unfortunately, long-distance DC power transmission has proven difficult and prohibitively expensive to implement in the "real world." Large-scale DC power transmission necessitates the use of massive, high-voltage AC-to-DC conversion systems at the generating plant, along with DC-to-AC power inverters at distribution stations where high-voltage lines branch out into lower-voltage, local lines.

PROBLEM 6-9

Mechanical power is sometimes expressed and measured in terms of units called *horsepower* (abbreviated HP). A mechanical power level of 1 HP equals approximately 746 W. Suppose that we need 2.01 HP of mechanical energy to turn a generator that puts out 1.23 kW of AC electricity. What's the efficiency of this generator as a ratio? As a percentage?

✔ SOLUTION

In this case $P_D = (746 \times 2.01)$ W, or 1.50 kW. We're told that $P_E = 1.23$ kW. The efficiency *Eff* as a ratio is

$$Eff = P_E/P_D$$
$$= 1.23/1.50$$
$$= 0.82$$

The efficiency $Eff_\%$ as a percentage equals 100 times the efficiency as a ratio. We can calculate

$$Eff_\% = 100\ (P_E/P_D)\%$$
$$= 100\ (1.23/1.50)\%$$
$$= (100 \times 0.82)\%$$
$$= 82\%$$

QUIZ

This is an "open book" quiz. You may refer to the text in this chapter. You'll find the correct answers listed in the back of the book.

1. Figure 6-7 shows a waveform as it might appear on a laboratory oscilloscope. Each vertical division represents 5.00 mV, and each horizontal division represents 100 ns. What's the positive peak voltage?
   A. +5.00 mV pk+
   B. +10.0 mV pk+
   C. +20.0 mV pk+
   D. +30.0 mV pk+

2. What's the negative peak voltage of the wave shown in Fig. 6-7?
   A. −2.00 mV pk−
   B. −5.00 mV pk−
   C. −10.0 mV pk−
   D. −20.0 mV pk−

3. What's the peak-to-peak voltage of the wave shown in Fig. 6-7?
   A. 6.00 mV pk-pk
   B. 20.0 mV pk-pk
   C. 30.0 mV pk-pk
   D. 60.0 mV pk-pk

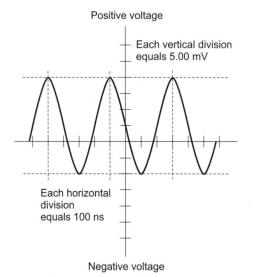

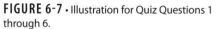

FIGURE 6-7 • Illustration for Quiz Questions 1 through 6.

4. Assuming that the waveform in Fig. 6-7 precisely follows the shape of a sinusoid, what's its average voltage?

   A. 0.00 mV avg
   B. +5.00 mV avg
   C. +10.0 mV avg
   D. +20.0 mV avg

5. What's the period of the waveform shown in Fig. 6-7?

   A. 400 ns
   B. 200 ns
   C. 100 ns
   D. 40.0 ns

6. What's the frequency of the waveform shown in Fig. 6-7?

   A. 400 kHz
   B. 625 kHz
   C. 1.25 MHz
   D. 2.50 MHz

7. The period of an AC wave varies inversely in proportion to its

   A. peak-to-peak amplitude.
   B. RMS amplitude.
   C. average amplitude.
   D. None of the above

8. The frequency of an AC wave varies inversely in proportion to its

   A. voltage.
   B. period.
   C. current.
   D. waveform.

9. If a generator produces 1500 W of useful electrical output power when we provide it with 2100 W of mechanical driving power, then its efficiency equals approximately

   A. 60.00%.
   B. 66.67%.
   C. 71.43%.
   D. 75.00%.

10. In the United States, a typical AC utility outlet provides 117 V RMS with no DC component. What's the peak-to-peak voltage?

    A. 331 V pk-pk
    B. 234 V pk-pk
    C. 165 V pk-pk
    D. 82.7 V pk-pk

chapter 7

Electricity in the Home

In this chapter, we'll look at some of the most important characteristics of standard utility AC. It's the form of electricity used in most homes and businesses.

CHAPTER OBJECTIVES

In this chapter, you will

- Learn how to break an AC wave down into degrees of phase.

- See how waves in differing phase combine.

- Compare single-phase and three-phase AC electricity.

- Learn how transformers can increase or decrease AC voltage.

- Calculate parameters using Ohm's Law for AC.

- Calculate AC power and energy values.

- Discover what causes lightning, and learn how to protect yourself and your hardware from its effects.

Phase

Phase allows us to express or define points in time during an AC cycle. Phase can also define or express time displacement between two AC waves. When we compare two waves, we can define their *phase difference* if, but only if, the waves have identical frequencies (and therefore identical periods).

Degrees of Phase

We can specify time points in an AC wave by dividing one complete cycle into 360 equal parts called *degrees* or *degrees of phase*. We assign 0 degrees (0°) to the point in the cycle where the magnitude is zero and positive-going. One-quarter of the way through the cycle corresponds to 90°, halfway through the cycle corresponds to 180°, three-quarters of the way through the cycle corresponds 270°, and the end of the cycle corresponds to 360°. Figure 7-1 illustrates this concept for a sine wave.

Waves in Phase Coincidence

We can define the *phase relationship, relative phase, phase difference,* or *phase angle* (all four terms mean the same thing) between two waves only when the two waves occur at the same frequency. If the frequencies differ, the relative phase constantly changes, so we can't obtain a definite value for the phase

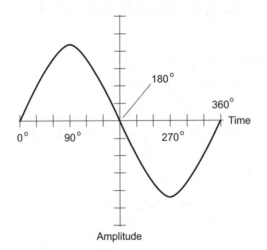

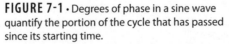

FIGURE 7-1 • Degrees of phase in a sine wave quantify the portion of the cycle that has passed since its starting time.

relationship. In the following discussions of relative phase, let's assume that the two waves always have the same frequency.

The term *phase coincidence* means that two waves having identical frequencies begin at exactly the same instant in time. If the waveforms have identical shape (although perhaps different amplitudes), they follow each other along from instant to instant. Figure 7-2A shows an example of phase coincidence between two waves whose amplitudes differ. The phase difference, or phase angle, equals 0°. In this situation, we can observe the following facts:

- The positive peak amplitude of the resultant wave, which is also a sine wave, equals the sum of the positive peak amplitudes of the two composite waves.
- The negative peak amplitude of the resultant wave equals the sum of the negative peak amplitudes of the composite waves.
- The peak-to-peak amplitude of the resultant wave equals the sum of the peak-to-peak amplitudes of the composite waves.
- The phase of the resultant wave coincides with the phases of the two composite waves.

Waves of Differing Phase

Two perfect sine waves having the same frequency can differ in phase by any amount from 0° (phase coincidence), through 90° (*phase quadrature*, meaning

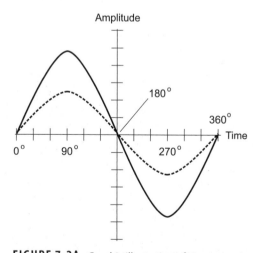

FIGURE 7-2A • Graphic illustration of sine waves in phase coincidence; the phase difference in this case equals 0°.

a difference of 1/4 cycle), through 180° (a difference of 1/2 cycle), through 270° (phase quadrature again, but a difference of 3/4 cycle), and finally 360° (phase coincidence, but offset by a full cycle).

Waves 180° out of Phase

When two pure sine waves of identical frequency begin exactly 1/2 cycle apart in time, we say that they occur *180° out of phase* with respect to each other. Figure 7-2B illustrates a situation of this sort.

If two sine waves have *identical* amplitudes and exist 180° out of phase, and if neither of them has a DC component, then they completely cancel each other out because the instantaneous amplitudes of the two waves are equal and opposite at every point in time. As a result, we get no output signal at all! If two sine waves have *different* amplitudes, neither of them has a DC component, and they occur 180° out of phase, then we can observe the following facts:

- The peak-to-peak amplitude of the resultant wave, which is also a sine wave, equals the difference between the peak-to-peak amplitudes of the two composite waves.
- The phase of the resultant wave coincides with the phase of the stronger of the two composite waves.
- The resultant wave has no DC component.

A sine wave has the unique property that, if we shift its phase by 180°, we'll obtain the same result as we get if we *invert* the original wave (we "turn it

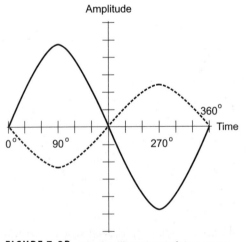

FIGURE 7-2B • Graphic illustration of sine waves 180° out of phase with each other.

upside-down"). A perfect, symmetrical square wave with no DC component also has this property. However, most other waveforms don't behave this way. We *cannot*, in general, say that a 180° phase shift is equivalent to *phase opposition* (inverting the waveform).

? Still Struggling

Theoretically, moving a wave forward or backward by 1/2 cycle doesn't change it in the same way as inverting it does. A pure sine wave with no DC component, or a symmetrical square wave with no DC component, constitute special cases where the two actions give us the same practical result in ordinary electrical circuits. However, with other waves, the results usually differ. To see these effects, graph some examples of non-sine waves or waves with DC components, and then compare the results of a 180º phase shift with wave inversion.

Single-Phase AC

Single-phase AC consists of a lone, single, pure sine wave, as shown in Fig. 7-1. You'll find this sort of AC at standard wall outlets intended for small appliances, such as lamps, TV sets, and computers. In most parts of the United States, the RMS voltage is standardized at 117 V, but it can vary a few percent above or below this level, depending on the overall utility power demand at the time, your location, and the whims of your local electric utility company. The positive and negative peak voltages equal 1.414 times the RMS voltage, or about +165 V pk+ and −165 V pk−. The peak-to-peak voltage equals about 2.828 times the RMS voltage, or 331 V pk-pk.

Three-Phase AC

Over long-distance power lines, utility companies usually transmit electricity in the form of three sine waves, each having the same RMS voltage, but differing in relative phase by 120° (1/3 of a cycle). Engineers call it *three-phase* AC. Figure 7-3 shows what a three-phase AC wave combination looks like if we plot it as a graph of instantaneous voltage versus time. We call the three individual waves *phase 1*, *phase 2*, and *phase 3*. The horizontal axis in this figure is graduated in degrees for phase 1 (solid curve). Phase 2 (coarse dashed curve) comes 1/3 of a

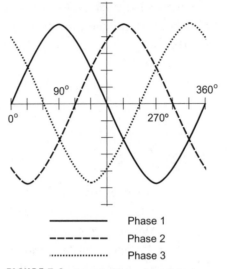

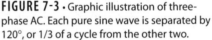

Phase 1
Phase 2
Phase 3

FIGURE 7-3 · Graphic illustration of three-phase AC. Each pure sine wave is separated by 120°, or 1/3 of a cycle from the other two.

cycle later than phase 1, and phase 3 comes 2/3 of a cycle later than phase 1. Alternatively, we can say that phase 2 comes 1/3 of a cycle earlier than phase 1.

TIP *Three-phase AC propagates (travels) more efficiently along power lines than single-phase AC propagates. As a result, less energy goes to waste as heat in the wires.*

PROBLEM 7-1

Suppose that two perfect AC sine waves, called wave X and wave Y, have the same frequency and occur in phase coincidence. Neither wave has any superimposed DC component. The peak-to-peak voltage of wave X equals 100 V pk-pk, and the peak-to-peak voltage of wave Y equals 150 V pk-pk. What's the peak-to-peak voltage of the composite wave Z, which we can define as X + Y? What's the positive peak voltage of Z? The negative peak voltage? The RMS voltage?

SOLUTION

The peak-to-peak voltage of wave Z equals the sum of the peak-to-peak voltages of waves X and Y. Therefore, we can calculate

$$E_{\text{Z-pk-pk}} = E_{\text{X-pk-pk}} + E_{\text{Y-pk-pk}}$$
$$= 100 + 150$$
$$= 250 \text{ V pk-pk}$$

Because neither wave has any DC component, the positive peak voltage of wave Z equals half the peak-to-peak voltage of wave Z. Therefore, we have

$$E_{Z\text{-pk}+} = +(E_{Z\text{-pk-pk}})/2$$
$$= +250/2$$
$$= +125 \text{ V pk}+$$

The negative peak voltage of wave Z equals half the peak-to-peak voltage of wave Z multiplied by −1, so we have

$$E_{Z\text{-pk-}} = -(E_{Z\text{-pk-pk}})/2$$
$$= -250/2$$
$$= -125 \text{ V pk}-$$

The RMS voltage of wave Z equals approximately 0.3536 times the peak-to-peak voltage of wave Z, so we get

$$E_{Z\text{-RMS}} = 0.3536 \, E_{Z\text{-pk-pk}}$$
$$= 0.3536 \times 250$$
$$= 88.4 \text{ V RMS}$$

PROBLEM 7-2

Imagine that two perfect AC sine waves called X and Y occur 180° out of phase. No DC component exists in either wave. The peak-to-peak voltage of wave X equals 100 V, and the peak-to-peak voltage of wave Y equals 150 V. What's the peak-to-peak voltage of the composite wave, Z = X + Y? The positive peak voltage? The negative peak voltage? The RMS voltage?

SOLUTION

When we have two perfect AC sine waves 180° out of phase, the peak-to-peak voltage of wave Z equals the *absolute value* of the difference between the peak-to-peak voltages of waves X and Y. From our prealgebra mathematics courses, we recall that if a number is zero or positive, then its absolute value equals its actual value; if a number is negative, then its absolute value equals −1 times its actual value. We can symbolize the absolute value of a quantity by placing vertical lines on either side of that quantity. There-fore, in this problem's scenario, we have

$$E_{Z\text{-pk-pk}} = | E_{X\text{-pk-pk}} - E_{Y\text{-pk-pk}} |$$
$$= | 100 - 150 |$$
$$= | -50.0 |$$
$$= -1 \times (-50.0)$$
$$= 50.0 \text{ V pk-pk}$$

The positive peak voltage of wave Z equals half of its peak-to-peak voltage. Therefore

$$E_{Z\text{-pk+}} = +(E_{Z\text{-pk-pk}})/2$$
$$= +50.0/2$$
$$= +25.0 \text{ V pk+}$$

The negative peak voltage of wave Z equals −1 times half of its peak-to-peak voltage. Therefore

$$E_{Z\text{-pk-}} = -(E_{Z\text{-pk-pk}})/2$$
$$= -50.0/2$$
$$= -25.0 \text{ V pk−}$$

The RMS voltage of wave Z equals approximately 0.3536 times the peak-to-peak voltage, so

$$E_{Z\text{-RMS}} = 0.3536 \, E_{Z\text{-pk-pk}}$$
$$= 0.3536 \times 50.0$$
$$= 17.7 \text{ V RMS}$$

Transformers

A *transformer* allows us to convert, or *transform*, an AC sine wave with a given RMS voltage to another AC sine wave with the same frequency but a different RMS voltage. Transformers take advantage of a phenomenon called *inductive coupling*, where fluctuating currents in a wire *induce* fluctuating currents at the same frequency in other wires nearby, even if those other wires have no direct connection to the main wire. In AC electrical applications, most trans-formers consist of wires wound on special molded forms called *cores* made of *laminated iron* (thin slabs of iron glued together) to maximize the amount of inductive coupling.

Windings

The *primary* of a transformer is the winding to which we apply the electricity whose voltage we want to change. The *secondary* is the winding from which we take the electricity after its voltage has changed. The nature and extent of the voltage transformation depends on the relative numbers of turns in the two windings.

1. In a *step-down transformer,* the primary has more turns than the secondary. The voltage across the primary (the input voltage) exceeds the voltage across the secondary (the output voltage).

2. In a *step-up transformer,* the primary has fewer turns than the secondary. The voltage across the secondary (the output voltage) exceeds the voltage across the primary (the input voltage).

Figure 7-4A shows the schematic symbol for a step-down transformer, and Fig. 7-4B shows the schematic symbol for a step-up transformer. The solid, parallel, vertical lines indicate that the transformers have laminated iron cores.

The optimum size, or *gauge,* of the wire in the transformer primary winding depends on the amount of current that we expect it to carry. As the current

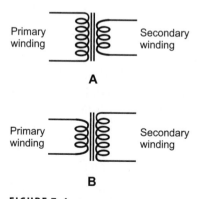

Primary winding   Secondary winding

A

Primary winding   Secondary winding

B

FIGURE 7-4 · At A, schematic symbol for a step-down AC transformer, where the primary has more turns than the secondary. At B, schematic symbol for a step-up AC transformer, where the secondary has more turns than the primary.

demand increases, so does the required wire diameter. For a given amount of power, a step-up transformer must use heavier-gauge primary-winding wire than a step-down transformer needs. A transformer that steps up the voltage by a large factor must have a primary comprising very heavy (large-diameter) wire.

The optimum gauge of the wire in the secondary winding of a transformer depends on the power demand and also on the *turns ratio*, which we define as the number of turns in the primary divided by the number of turns in the sec-ondary. A step-up transformer (turns ratio less than 1:1) can have thinner secondary-winding wire than a step-down transformer (turns ratio larger than 1:1) for a given amount of power.

TIP   *The secondary winding of a transformer can have intermediate connections, called* taps, *for obtaining various AC output voltages.*

Operation

Small transformers are used in *power supplies* for electronic devices, such as computers and radios. We'll learn how power supplies work in the next chapter. Medium-sized transformers find application in high-current or high-voltage power supplies. Large transformers provide the utility power that we consume in our homes and businesses. The bulkiest and most massive transformers, some as large and heavy as farm tractors or dump trucks, can handle thousands of volts and thousands of amperes; they serve power-transmission stations.

If a transformer has a primary winding with N_{pri} turns and a secondary wind-ing with N_{sec} turns, then we define the primary-to-secondary turns ratio T as

$$T = N_{pri}/N_{sec}$$

For a transformer with a primary-to-secondary turns ratio of T and an effi-ciency of 100% (no power loss), the RMS voltage E_{sec} across the entire second-ary winding is related to the RMS voltage E_{pri} across the entire primary winding according to the following equations:

$$T = E_{pri}/E_{sec}$$

$$E_{pri} = E_{sec} T$$

$$E_{sec} = E_{pri}/T$$

Based on the foregoing formulas, we can use algebra to determine that

$$E_{pri}/E_{sec} = N_{pri}/N_{sec}$$

Efficiency

In a "real-world" transformer, some power always gets lost in the coil windings. Some power also gets lost in the core. *Conductor losses* occur because of the *ohmic resistance* of the wire that makes up the windings. *Core losses* occur because of *eddy currents* (circulating currents in the iron) and *hysteresis* ("sluggishness" of the iron's response to AC magnetic fields). The efficiency of a transformer is, therefore, always less than 100%. The power loss appears in the form of heat; the windings and the core rise in temperature as the transformer does its job.

Let E_{pri} and I_{pri} represent the RMS primary-winding voltage and current in a hypothetical transformer, and let E_{sec} and I_{sec} represent the RMS secondary-winding voltage and current. In a perfect transformer, the product $E_{pri} I_{pri}$ would equal the product $E_{sec} I_{sec}$. However, in a "real-world" transformer, $E_{pri} I_{pri}$ always exceeds $E_{sec} I_{sec}$. We can calculate the efficiency *Eff* of a transformer as the ratio

$$Eff = E_{sec} I_{sec} / (E_{pri} I_{pri})$$

Expressed as a percentage, the efficiency $Eff_\%$ is

$$Eff_\% = 100\, E_{sec} I_{sec} / (E_{pri} I_{pri})$$

The power P_{loss} dissipated in the transformer windings and core, and thereby wasted as heat, is

$$P_{loss} = E_{pri} I_{pri} - E_{sec} I_{sec}$$

These formulas will work for electrical units of volts RMS (V RMS), amperes RMS (A RMS), and watts RMS (W RMS), as long as no DC component exists in the input or output, and as long as we don't connect anything to the transformer output that shifts the wave phase.

If a DC component does exist in either the input or the output of a transformer, or if we connect the secondary winding to an appliance that shifts the phase of the AC wave, then the situation becomes rather complicated, and most of the above formulas no longer hold strictly true. However, in a well-engineered transformer circuit with DC components, we still observe the characteristic relationship between peak-to-peak voltages and the turns ratio. If $E_{pk-pk-pri}$ represents the peak-to-peak voltage across the primary and $E_{pk-pk-sec}$ represents the peak-to-peak voltage across the secondary, then

$$E_{pk-pk-pri} / E_{pk-pk-sec} = N_{pri} / N_{sec}$$

TIP   *An AC wave can pass through a transformer, but DC can't. If the input wave to a transformer contains a DC component, that component fails to appear in the output. Conversely, if we impose a DC component in the secondary circuit of a transformer, that component won't get transferred to the primary circuit. Engineers sometimes take advantage of this "DC-blocking" property when they use transformers in specialized electronic devices, such as signal amplifiers.*

TIP   *The efficiency of a transformer varies depending on the load connected to the secondary winding. If the secondary-circuit current drain rises too high, the transformer efficiency goes down. The efficiency will also suffer if a DC component exists in the input wave. Transformers are rated according to the maximum amount of output power that they can deliver without serious degradation in efficiency.*

PROBLEM 7-3

Suppose that the AC input voltage to a step-down transformer equals 120 V RMS, constitutes a pure sine wave, and contains no DC component. This voltage E_{pri} appears across the entire primary winding. Suppose that the primary-to-secondary turns ratio, T, equals exactly 5:1. The transformer has a secondary with a tap in the center (Fig. 7-5). What's the RMS output voltage E_{sec} that appears across the entire secondary winding between terminals X and Z?

SOLUTION

The RMS secondary voltage equals exactly 1/5 of the RMS primary voltage. This fact becomes apparent when we plug numbers into the formula. Calculating, we get

$$E_{sec} = E_{pri}/T$$
$$= 120/5.00$$
$$= 24.0 \text{ V RMS}$$

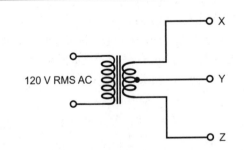

FIGURE 7-5 · Illustration for Problems 7-3 and 7-4.

PROBLEM 7-4

What's the RMS voltage that appears between terminals X and Y of the secondary winding in the circuit of Fig. 7-5, assuming the same turns ratio as in Problem 7-3? What's the RMS voltage that appears between terminals Y and Z? As before, assume that the input is a sine wave with no DC component. Also remember that the tap, at terminal Y, connects to the center of the secondary winding.

SOLUTION

The tap splits the secondary voltage into two equal parts. The voltage between terminals X and Y, therefore, equals (24.0/2.00) or 12.0 V RMS, and the voltage between terminals Y and Z also equals (24.0/2.00) or 12.0 V RMS. Alternatively, we can say that the transformer has "two secondaries," one between terminals X and Y, and the other between terminals Y and Z. In that case, both "sub-transformers" have turns ratios of exactly 10:1, and the RMS output voltage across either "sub-secondary", therefore, equals 120/10.0, or 12.0 V RMS.

PROBLEM 7-5

Suppose that the AC input voltage to a step-up transformer equals 12.0 V RMS. This voltage E_{pri} appears across the entire primary. Suppose that the primary-to-secondary turns ratio T equals precisely 1:8, which works out to a value of 0.125. What's the *peak-to-peak* voltage $E_{pk-pk-sec}$ that appears across the entire secondary winding? Assume that the input is a sine wave with no DC component.

SOLUTION

The RMS (not peak-to-peak) secondary voltage E_{sec} equals exactly eight times the RMS primary voltage. That's 12.0 × 8.00 or 96.0 V RMS. This fact becomes apparent when we plug numbers into the appropriate formula, getting

$$E_{sec} = E_{pri}/T$$
$$= 12.0/0.125$$
$$= 12.0 \times 8.00$$
$$= 96.0 \text{ V RMS}$$

The peak-to-peak voltage $E_{pk-pk-sec}$ across the secondary equals the RMS voltage times 2.828. Calculating, we obtain

$$E_{pk-pk-sec} = 2.828\, E_{sec}$$
$$= 96.0 \times 2.828$$
$$= 271 \text{ V pk-pk}$$

PROBLEM 7-6

Imagine that we connect a simple load across the secondary winding of a transformer. Suppose that the voltage across the primary winding equals 120 V RMS, and the current through the primary winding equals 2.57 A RMS. Further suppose that the voltage across the secondary winding equals 12.3 V RMS, and the current drawn by the load connected across the secondary winding equals 19.9 A RMS. What's the efficiency of this transformer, expressed as a ratio? As a percentage?

✔ SOLUTION

To determine the efficiency, we plug in the numbers to the formula and then calculate, getting

$$Eff = E_{sec}\, I_{sec} / (E_{pri}\, I_{pri})$$
$$= (12.3 \times 19.9) / (120 \times 2.57)$$
$$= 244.77 / 308.4$$
$$= 0.794$$

To express the efficiency as a percentage, we multiply Eff by 100, obtaining the figure $Eff_\% = 79.4\%$.

Ohm's Law, Power, and Energy

With respect to Ohm's Law, power, and energy, household utility AC behaves like DC as long as we always use RMS AC values, and as long as the current and the voltage remain in phase with each other.

Reactance and Impedance

When the current and the voltage in an AC circuit don't follow each other in phase, quantities known as *reactance* and *complex impedance* replace ordinary resistance. An analysis of these phenomena goes beyond the scope of this course. If you're interested in learning more about them, I recommend *Teach*

Yourself Electricity and Electronics (comprehensive with lots of mathematics) or *Electronics Demystified* (more concise but less mathematical).

Ohm's Law for AC

Imagine a circuit containing a source of AC electricity, a load (shown as a resistor), and an AC ammeter that measures the RMS current through the resistor (Fig. 7-6). Let E_{RMS} stand for the voltage of the AC source (in volts RMS), let I_{RMS} stand for the current through the load (in amperes RMS), and let R stand for the resistance of the load (in ohms). Three formulas denote Ohm's Law for AC:

$$E_{RMS} = I_{RMS}\,R$$
$$I_{RMS} = E_{RMS}/R$$
$$R = E_{RMS}/I_{RMS}$$

Still Struggling

You must always use units of volts RMS (V RMS), amperes RMS (A RMS), and ohms if you want the above-stated formulas to work right. If quantities are given in units other than volts RMS, amperes RMS, and ohms, you must convert everything to those units and then do your calculations. After that, you can express your final answer in whatever unit you like.

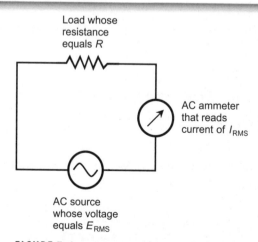

Load whose resistance equals R

AC ammeter that reads current of I_{RMS}

AC source whose voltage equals E_{RMS}

FIGURE 7-6 · A basic AC circuit for demonstrating Ohm's Law, power, and energy calculations. Illustration for Problems 7-7 through 7-15.

PROBLEM 7-7

Suppose that the AC source in the circuit of Fig. 7-6 produces 40 V RMS, and the resistor has a value of 20 ohms. What's the RMS current?

✔ SOLUTION

Plug in the numbers to the Ohm's Law formula for current to obtain

$$I_{RMS} = E_{RMS}/R$$
$$= (40/20)$$
$$= 2.0 \text{ A RMS}$$

PROBLEM 7-8

Suppose that the load in the circuit of Fig. 7-6 has a resistance of 100 ohms, and the measured current equals 100 mA RMS. What's the RMS voltage of the AC source?

✔ SOLUTION

First, we convert the current to amperes, getting $I_{RMS} = 0.100$ A RMS. Then we plug in the numbers and calculate, obtaining

$$E_{RMS} = I_{RMS} R$$
$$= 0.100 \times 100$$
$$= 10.0 \text{ V RMS}$$

PROBLEM 7-9

If the AC source voltage in the circuit of Fig. 7-6 is 117 V RMS and the ammeter shows 3.0 A RMS, what's the value of the resistor?

✔ SOLUTION

We can plug in the values directly, because they're expressed in volts RMS and amperes RMS. We obtain

$$R = E_{RMS}/I_{RMS}$$
$$= (117/3.0) \text{ ohms}$$
$$= 39 \text{ ohms}$$

AC Power Calculations

Three formulas allow us to determine the AC power, P (in watts) consumed by the load in the generic circuit of Fig. 7-6:

$$P = E_{RMS} \, I_{RMS}$$

$$P = (E_{RMS})^2 / R$$

$$P = (I_{RMS})^2 \, R$$

? Still Struggling

You must always use units of volts RMS (V RMS), amperes RMS (A RMS), and ohms for power calculations if you want the above-stated formulas to come out in watts. If any of the quantities is specified in some unit other than these, you must convert before you calculate.

PROBLEM 7-10

Suppose that the AC source voltage in the circuit of Fig. 7-6 is 60.0 V RMS and the ammeter reads 5.00 A RMS. How much power does the load consume?

✔ SOLUTION

We can input in the numbers directly to the formula for power in terms of voltage and current, calculating to obtain

$$P = E_{RMS} \, I_{RMS}$$
$$= 60.0 \times 5.00$$
$$= 300 \, W$$

PROBLEM 7-11

Suppose that the AC source voltage in the circuit of Fig. 7-6 produces 10 V RMS, and the load has a resistance of 50 ohms. How much power does the load consume?

SOLUTION

We can input the values directly into the formula for power in terms of voltage and resistance, and calculate to get

$$P = (E_{RMS})^2 / R$$
$$= 10 \times 10 / 50$$
$$= 2.0\ W$$

PROBLEM 7-12

Suppose that the load in the circuit of Fig. 7-6 has an internal resistance of 9.73 ohms, and we measure the current through it as 6.11 A RMS. What's the power?

SOLUTION

We can use the formula for power in terms of current and resistance. When we plug in the numbers and calculate, we get

$$E = (I_{RMS})^2 R$$
$$= 6.11 \times 6.11 \times 9.73$$
$$= 363\ W$$

AC Energy Calculations

We can use any of the following three formulas to determine the energy Q (in watt-hours) consumed by the load in the circuit of Fig. 7-6:

$$Q = E_{RMS}\ I_{RMS}\ t$$
$$Q = (E_{RMS})^2\ t / R$$
$$Q = (I_{RMS})^2\ R\ t$$

? Still Struggling

You must always use units of volts RMS (V RMS), amperes RMS (A RMS), ohms, and hours (h) for energy calculations to come out in watt-hours. If quantities are given in units other than the standard ones, you must convert all the units before you plug any numbers into any of the foregoing formulas.

PROBLEM 7-13

Suppose that we activate the circuit of Fig. 7-6 for 3.00 h and then switch it off. Imagine that the AC voltage source supplies 13.5 V RMS and the ammeter shows 4.25 A. How much energy, in watt-hours, does the resistor consume?

SOLUTION

We use the first of the three formulas above. We're given the quantities in standard units, so we don't have to convert any of them. Here, $E_{RMS} = 13.5$, $I_{RMS} = 4.25$, and $t = 3.00$. We calculate to get

$$Q = E_{RMS} I_{RMS} t$$
$$= 13.5 \times 4.25 \times 3.00$$
$$= 172\, Wh$$

PROBLEM 7-14

Imagine that we don't know the ammeter reading in the circuit of Fig. 7-6, but we know that the circuit remains activated for 15.0 min, and we know that the AC source voltage equals 240 V RMS. We also know that the load resistance equals 62.5 ohms. How much energy, in watt-hours and kilowatt-hours, does the load consume during this 15.0-min period of time?

SOLUTION

First, let's convert the time to hours, getting 15.0 min = 0.250 h. Therefore, $t = 0.250$. Then we should note that $E_{RMS} = 240$ and $R = 62.5$. Now we can use the second of the above formulas for energy to obtain

$$Q = (E_{RMS})^2\, t / R$$
$$= 240 \times 240 \times 0.250 / 62.5$$
$$= 230\, Wh$$
$$= 0.230\, kWh$$

PROBLEM 7-15

Suppose that we don't know the AC source voltage in the circuit shown in Fig. 7-6, but we know that the load has a resistance of 200 ohms. The circuit is activated for 90.0 min, and during this time, the ammeter reads 500 mA RMS.

How much energy, in watt-hours and kilowatt-hours, does the load consume during this 90.0-min period of time?

 SOLUTION

First, we must convert the time to hours: 90.0 min = 1.50 h. Therefore, $t = 1.50$. Then we convert the current to amperes: 500 mA = 0.500 A. Therefore, $I = 0.500$. We also note that $R = 200$. When we plug these numbers into the third formula for energy given above, we get

$$Q = (I_{RMS})^2 R t$$
$$= 0.500 \times 0.500 \times 200 \times 1.50$$
$$= 75.0 \text{ Wh}$$
$$= 0.0750 \text{ kWh}$$

Lightning

No discussion of home electrical systems would do justice to the topic without mentioning a phenomenon that everyone experiences sooner or later: *lightning*.

Causes

A constant voltage exists between the earth's surface and the portion of the upper atmosphere called the *ionosphere*, where sun-induced *ionization* of gases causes some conduction to occur. The lower atmosphere acts as a *dielectric* (insulator), so the earth and atmosphere in effect form a giant *capacitor* (electrical storage system). This so-called *atmospheric capacitor* acquires a stupendous quantity of electrical charge. The insulating lower atmosphere frequently breaks down under the stress of the resulting potential difference. That's when lightning occurs.

Lightning discharges tend to occur most often in and near areas of precipitation, particularly in rainstorms. Lightning occurs in tropical storms and hurricanes, although some such systems have a lot more lightning than others do. Lightning can take place in heavy snow squalls and blizzards (especially mountain storms), and it can even strike once in awhile from a partly cloudy sky while the sun shines and no precipitation reaches the surface. Sandstorms and erupting volcanoes can also produce lightning.

TIP   *Whenever a potential difference builds up indefinitely between two points or regions, a lightning discharge will eventually take place. Lightning's capricious* *nature has spawned legends and myths dating back to the beginning of civilization. Some of these tales have a basis in fact; others have been embellished by imaginative storytellers to the point of fantasy. One fact remains certain: Lightning presents a deadly danger.*

The Stroke

A lightning "bolt" is technically called a *stroke*. It lasts for only a small fraction of a second, but in this time, it can start fires, cause explosions, destroy electrical and electronic equipment, and electrocute people and animals. Four types of stroke can occur:

1. A discharge that takes place within a single cloud (an *intracloud stroke*)
2. A discharge in which the electrons travel from a cloud to the earth (a *cloud-to-ground stroke*)
3. A discharge in which the electrons travel from one cloud to another (an *intercloud stroke*)
4. A discharge in which the electrons travel from the earth to a cloud (a *ground-to-cloud stroke*)

Meteorologists and engineers usually consider lightning-stroke current to flow in the same direction as the electrons move: from the negative charge pole to the positive charge pole. This expression runs contrary to the physicist's definition of conventional current from positive to negative.

A lightning stroke begins when the charge quantity between two points or regions gets so large that electrons begin to advance through the air, moving away from the negative charge pole. A small current, called a *stepped leader*, "seeks out and finds" the path of least resistance through the atmosphere. Once the stepped leader has established this route (usually within a fraction of a second), some of the air molecules along the path become *ionized* (electrically charged). Engineers call the ionized path, which typically has a jagged nature, the *channel*. When the stepped leader has fully established a conductive channel through the atmosphere from the negative charge pole to the positive charge pole, we observe one or more *return strokes*, attended by a massive *current surge* that can peak at over 100,000 A.

TIP   *The direct destructive effects of lightning result from energy dissipated when the current forces its way through objects, such as trees, buildings, aircraft, living things, or anything else in its way.*

Personal Protection

In the United States, lightning kills more people than hurricanes or tornadoes. Burns and electrocution constitute the usual causes of death. Property damage can result from fires, induced currents, and explosions. People can suffer real and lasting harm from lightning; electronic equipment can undergo damage or destruction despite all reasonable precautions. Nevertheless, you can take certain steps to keep the danger to a minimum. The following precautions can minimize personal risk (but can't guarantee absolute immunity):

- Stay indoors, or inside an insulated metal enclosure, such as a car, bus, or train.
- Stay away from windows.
- If you can't get indoors, find a low spot on the ground, such as a ditch or ravine, and squat down with your feet close together until the threat has passed.
- Avoid lone trees or other isolated, tall objects, such as utility poles or flagpoles.
- Avoid electrical appliances or electronic equipment connected to utility power lines or to an external antenna.
- Stay out of the shower or bathtub.
- Avoid swimming pools, whether indoors or outdoors.
- Do not use hard-wired telephone sets.

Protecting Hardware

Precautions that minimize the risk of damage to electrical and electronic equipment (but do not guarantee immunity) include the following:

- Never operate, or experiment with, a hobby radio station when lightning occurs near your location.
- Disconnect and ground all radio, television, or communications antennas.
- Unplug all sensitive appliances from wall outlets when a thunderstorm approaches.
- Devices called *lightning arrestors* provide some protection from electrostatic-charge buildup, but they cannot offer complete safety, and you should not rely on them for routine protection.

- One or more well-grounded *lightning rods* can reduce (but not eliminate) the chance of a direct hit, but you should never use them as an excuse to neglect the other precautions.

- Power line *transient suppressors* (also called "surge protectors") reduce computer "glitches" and can sometimes protect sensitive components, but you shouldn't use them as an excuse to neglect the other precautions.

- You can find other secondary protection devices advertised in electronics-related magazines and on Web sites.

TIP   *If you can hear thunder, then make no mistake: Lightning exists nearby whether you can see it or not, and it's close enough to present a genuine peril.*

TIP   *If you see a lightning stroke and then hear the resulting thunder, count the number of seconds between the flash and the first audible noise. Divide seconds by 5 to determine the distance in miles between you and the stroke. Divide seconds by 3 to determine the distance in kilometers between you and the stroke.*

?

Still Struggling

To learn more about how to protect your hardware and yourself against the hazards of lightning, consult a competent engineer or electrician. If you want to ascertain the fire safety level of your house or business, consult your local fire inspector.

QUIZ

This is an "open book" quiz. You may refer to the text in this chapter. You'll find the correct answers listed in the back of the book.

1. A transformer is designed to step 234 V RMS utility AC down to 117 V RMS AC for use with common appliances. The transformer primary has 500 turns. How many turns does the secondary have?

    A. 2000 turns
    B. 1000 turns
    C. 250 turns
    D. 125 turns

2. A phase shift of 120° represents

    A. 1/6 of a cycle.
    B. 1/4 of a cycle.
    C. 1/3 of a cycle.
    D. 2/3 of a cycle.

3. Suppose that a certain load constitutes a pure resistance of 47 ohms. We measure an AC current through it as 335 mA RMS. How much power does the load dissipate?

    A. 5.3 W
    B. 16 W
    C. 140 W
    D. We need more information to calculate it.

4. Suppose that a certain load constitutes a pure resistance of 100 ohms. We measure an AC voltage across it as 10.0 V RMS. How much power does the load dissipate?

    A. 100 mW
    B. 1.00 W
    C. 10.0 W
    D. 100 W

5. Imagine a load that offers a constant, pure resistance to AC. We connect the load to a constant, pure-sine-wave AC voltage source of 100 V RMS for a few minutes, causing the load to dissipate a certain amount of energy. Then we repeat the experiment for the same period of time, but using a larger constant, pure-sine-wave AC voltage. We want the load to dissipate twice as much energy as it did the first time. What RMS voltage should we use?

    A. 141 V RMS
    B. 200 V RMS
    C. 400 V RMS
    D. 800 V RMS

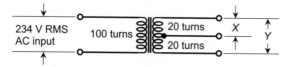

FIGURE 7-7 · Illustration for Quiz Question 8.

6. Consider two pure sine waves, neither of which has a DC component, and both of which occur at the same frequency. If one of the waves is offset by 1/2 cycle relative to the other, then in effect, the two waves exist in phase

    A. quadrature.

    B. coincidence.

    C. opposition.

    D. realignment.

7. If you see a flash of lightning and then you hear the resulting thunder 9 seconds later, you know that the lightning stroke occurred approximately

    A. 1 kilometer away.

    B. 3 kilometers away.

    C. 6 kilometers away.

    D. 9 kilometers away.

8. Figure 7-7 illustrates an AC transformer with a tap in the center of the secondary winding. The primary winding has exactly 100 turns. The entire secondary has exactly 40 turns; the tap divides it into two sections of exactly 20 turns each. We connect a constant, pure sine-wave AC source of 234 V RMS to the primary winding. What AC voltages X and Y should we expect to observe?

    A. $X = 9.36$ V RMS and $Y = 37.4$ V RMS

    B. $X = 9.36$ V RMS and $Y = 18.7$ V RMS

    C. $X = 46.8$ V RMS and $Y = 187$ V RMS

    D. $X = 46.8$ V RMS and $Y = 93.6$ V RMS

9. Suppose that two perfect AC sine waves, both having the same frequency, occur in phase coincidence. Neither wave has any superimposed DC component. One wave exhibits 25 V pk-pk, and the other wave exhibits 11 V pk-pk. The composite wave therefore has

    A. 11 V pk-pk.

    B. 14 V pk-pk.

    C. 25 V pk-pk.

    D. 36 V pk-pk.

10. In the scenario of Question 9, what's the approximate RMS voltage of the composite wave?

    A. 13 V RMS

    B. 26 V RMS

    C. 31 V RMS

    D. 40 V RMS

chapter 8

Electrical Power Supplies

An *electrical power supply* converts utility AC to DC as an alternative to batteries for electronic devices. Figure 8-1 illustrates the major components of a power supply that converts 117 V RMS AC to constant, relatively pure DC. As you read on, you'll learn the meanings of, and the actions represented by, the terms in the boxes.

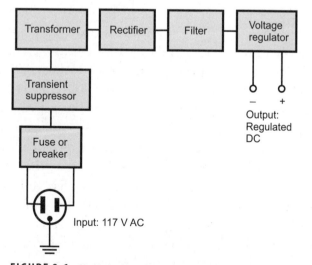

FIGURE 8-1 · Block diagram of a DC power supply. As you go through this chapter, you'll learn what the various boxes do.

CHAPTER OBJECTIVES

In this chapter, you will

- Understand diode action, forward bias, and reverse bias.
- Analyze and compare half-wave and full-wave rectifier circuits.
- Learn how to filter and regulate rectifier output to obtain pure DC.
- See how to protect appliances from transients and other anomalies.
- Learn how uninterruptible power supplies work.

Rectifiers

A *rectifier* converts AC to *pulsating DC*, usually by means of one or more heavy-duty *semiconductor diodes* following a power transformer. A diode passes current in only one direction. An applied voltage (known as a *bias voltage* or simply *bias*) that drives current through a diode in its "normal conduction direction" is called *forward bias*. An applied voltage of the opposite polarity, going against the "normal conduction direction," constitutes *reverse bias*. In all semiconductor diodes, it takes a certain minimum forward bias voltage to produce current. In a silicon diode (the most common type), this threshold, known as the *forward breakover voltage*, equals approximately 0.6 V. In other diode types it can range from approximately 0.3 V to around 2 V.

Half-Wave Circuit

The simplest type of rectifier circuit, called a *half-wave rectifier* (Fig. 8-2A), uses one diode (or a series or parallel combination of diodes if high voltage or current is required) to cut off half of the AC cycle. The effective voltage in this type of circuit equals roughly 35% of the positive or negative peak voltage, depending on the polarity of the pulsating DC output. Figure 8-3A shows a graphical example. The peak voltage in the reverse direction, called the *peak inverse voltage* (PIV) or *peak reverse voltage* (PRV) across the diode, can reach values of up to 2.8 times the applied RMS AC voltage. To provide an extra margin of safety, most engineers like to use diodes whose PIV ratings equal at least 1.5 times the maximum expected PIV. Therefore, when we build a half-wave power supply, we should use diodes rated for at least 2.8 × 1.5, or 4.2, times the RMS AC voltage that appears across the secondary winding of the transformer.

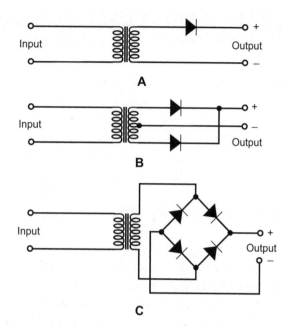

FIGURE 8-2 · At A, a half-wave rectifier circuit. At B, a full-wave center-tap rectifier circuit. At C, a full-wave bridge rectifier circuit.

Half-wave rectification has shortcomings. First, we might find it difficult to completely get rid of the pulsations in the output DC, especially if the load demands high current. Second, the output voltage can diminish considerably when the supply delivers high current. Third, half-wave rectification puts a

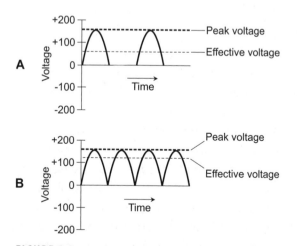

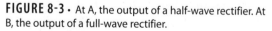

FIGURE 8-3 · At A, the output of a half-wave rectifier. At B, the output of a full-wave rectifier.

strain on the power transformer and the diodes because it *pumps* them. They can "loaf" during the half of the AC cycle when no current flows, but they must "strain" during the half of the cycle when current does flow.

Half-wave rectification usually works okay in a power supply that never has to deliver much current, or when the voltage can vary without affecting the behavior of the equipment connected to it. The main advantage of a half-wave circuit is its low cost compared with more sophisticated circuits.

? Still Struggling

Normally, electrons flow through a diode opposite the arrow in the diode's schematic symbol, but not in the direction that the arrow points. Conventional current flows in the same direction as the arrow points. Therefore, the arrow points from the more positive side of the diode to the more negative side. The arrow represents the diode's *anode*; the straight-line part of the symbol represents the *cathode*. The diode conducts current when the anode is positive with respect to the cathode to an extent that's equal to or greater than the forward breakover voltage.

TIP   *When we apply a voltage to a diode so that the anode is more positive than the cathode, we have forward bias. When we apply a voltage making the anode more negative than the cathode, we have reverse bias. If the anode and the cathode have the same electrical potential (no applied voltage difference exists between them), we have a condition called zero bias.*

Avalanche Effect

If we reverse-bias a diode, it won't conduct current—normally! But if we apply reverse bias and then "ratchet up" the reverse voltage without limit, we'll eventually reach a potential difference at which the diode suddenly begins to conduct in the reverse direction. Engineers call this phenomenon the *avalanche effect,* and the threshold voltage at which it begins the *avalanche voltage.* The avalanche voltage varies among different kinds of diodes. In a rectifier diode, the avalanche voltage is usually hundreds, and occasionally thousands, of times higher than the forward breakover voltage. When we design and build a power supply, we'll always want to make sure that the diodes have avalanche-voltage ratings high enough so that they won't leak any current at the reverse-bias voltage peaks in the AC cycle.

Full-Wave Center-Tap Circuit

We can take advantage of both halves of the AC cycle, rather than only the positive half or the negative half, to obtain pulsating DC. A *full-wave center-tap rectifier* has a transformer with a tapped secondary, as shown in Fig. 8-2B. The center tap connects directly to *electrical ground*, usually the metal *chassis* of the system, which in turn goes to the utility AC ground connection. The voltages and currents at the ends of the secondary winding occur in phase opposition with respect to each other. We individually half-wave rectify these two AC waves, alternately and repeatedly cutting off one half of the cycle and then the other.

In the circuit of Fig. 8-2B, the transformer center tap provides the negative DC output pole (that also happens to go to the electrical ground), while the diodes provide the positive voltage. We can reverse the polarity of the DC output by reversing the orientations of the diodes. The effective voltage equals approximately 71% of the positive or negative peak voltage, as shown in the graph of Fig. 8-3B. The PIV across the diodes can, as in the half-wave rectifier circuit, range up to 2.8 times the applied RMS AC voltage. Therefore, the diodes should have a PIV rating of at least 2.8×1.5, or 4.2, times the applied RMS AC voltage to ensure that they won't undergo avalanche breakdown at any time during the wave cycle.

TIP   *The "safety factor" of 1.5 (equal to an extra 50% of rated values) is a common standard in all fields of engineering. We might build a bridge to withstand 1500 tons, for example, even though we're certain that it will never have to hold a load of more than 1000 tons. We might build a wireless communications antenna to withstand winds of up to 120 miles per hour, even though our area has historically never experienced winds of more than 80 miles per hour. Once in awhile, an engineer will use a "safety factor" even greater than 50%; this practice is sometimes called* overengineering.

Full-Wave Bridge Circuit

In most applications, the *full-wave bridge rectifier*, sometimes called a *bridge* if we know the context, offers the best method for converting AC to DC in a power supply. Figure 8-2C shows a schematic diagram of a full-wave bridge circuit. The output waveform is identical to that produced by a full-wave center-tap circuit.

The effective output voltage from a full-wave bridge rectifier circuit equals about 71% of the peak voltage, as is the case with full-wave, center-tap rectification. The PIV across the diodes, however, equals only about 1.4 times the applied RMS AC voltage. Therefore, each diode needs to have a minimum PIV rating of only 1.4 × 1.5, or 2.1, times the RMS AC voltage that appears at the transformer secondary.

The bridge circuit doesn't need a center-tapped transformer secondary. The circuit takes advantage of the entire secondary winding on both halves of the wave cycle, so the bridge rectifier makes more efficient use of the transformer than the half-wave or the full-wave center-tap circuits do. The bridge circuit also treats the diodes more gently than either of the other rectifier arrangements do.

A bridge rectifier requires four diodes rather than two (in the case of a full-wave center-tap circuit) or one (in the case of a half-wave circuit). This increased complexity rarely amounts to much in terms of cost because most rectifier diodes are inexpensive, but it makes a big difference when a power supply must deliver high current. In that case, the extra diodes—two for each half of the cycle, rather than one—can dissipate more overall heat energy.

TIP *Engineers find it easier to eliminate the pulsations, called ripple, in the output DC from a full-wave rectifier than in the output from a half-wave rectifier. The full-wave rectifier also places less strain on the transformer and diodes than a half-wave circuit does. If we connect a significant load to the output of the full-wave rectifier circuit, forcing it to deliver high current, the voltage drops less than it would with a half-wave circuit under the same circumstances. In all these respects, a full-wave rectifier circuit outperforms a half-wave circuit designed for similar voltage and current.*

Voltage Multiplier

We can connect diodes and capacitors together to make a power supply deliver an output voltage at a whole-number multiple of the peak AC input voltage. Theoretically, we can get large whole-number voltage multiples, but we'll rarely see power supplies that make use of multiplication factors larger than 2.

In practice, *voltage-multiplier power supplies* work well only when the load draws low current. Otherwise, we end up having to contend with poor *voltage regulation,* meaning that the output voltage goes down *a lot* when the current demand rises significantly. In high-current, high-voltage applications, the best way to build a power supply involves the use of a step-up transformer, not a voltage-multiplier circuit.

Figure 8-4 shows a *voltage-doubler power supply*. This circuit works on the entire AC cycle, so engineers call it a *full-wave voltage doubler*. Its DC output

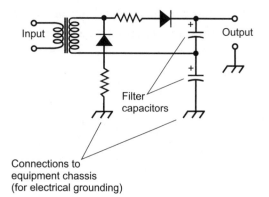

FIGURE 8-4 · A full-wave voltage-doubler power supply.

voltage, when the current drain is low, equals roughly twice the peak AC input voltage, or about 2.8 times the applied RMS AC voltage. A full-wave voltage doubler subjects the diodes to a PIV of 2.8 times the applied RMS AC voltage. Therefore, the diodes should have a PIV rating of at least 2.8 × 1.5, or 4.2, times the RMS AC voltage that appears across the transformer secondary.

TIP *Proper operation of a voltage-doubler or voltage-multiplier circuit depends on the ability of the capacitors to hold a charge under maximum load. Therefore, the capacitors must have large values, as well as be capable of handling high voltages. The capacitors serve two purposes: to boost the voltage and to get rid of the ripple in the DC output. The resistors, which have low ohmic values, protect the diodes against surge currents that occur when we "switch the supply on," that is, when we initially apply AC power to the transformer.*

▢ PROBLEM 8-1

Suppose that we connect a power transformer with a primary-to-secondary turns ratio of exactly 1:2 to the 117-V RMS AC utility mains. We use a half-wave rectifier circuit to obtain pulsating DC output. What's the minimum PIV rating the diodes should have in order to ensure that they won't break down? Round the answer off to the nearest volt.

✔ SOLUTION

The utility mains supply 117 V RMS AC, and the transformer steps up this voltage by a factor equal to the primary-to-secondary turns ratio. Therefore, the output of the transformer is 2 × 117 V RMS AC, or 234 V RMS AC. In a half wave circuit, the PIV rating of the diodes should equal at least 4.2 times that value, so they must be rated for at least 4.2 × 234 PIV, or 983 PIV.

PROBLEM 8-2

Suppose that we use a full-wave bridge rectifier circuit in the foregoing scenario, rather than a half-wave rectifier circuit. What minimum PIV rating must the diodes have in this case? Round the answer off to the nearest volt.

✔SOLUTION

The secondary still delivers 234 V RMS AC. In a full-wave bridge circuit, the PIV rating of the diodes should be at least 2.1 times the RMS AC voltage at the transformer secondary, or half the PIV in the half-wave situation: 2.1 × 234 PIV, or 491 PIV.

PROBLEM 8-3

Suppose that we use a voltage-doubler circuit with the transformer described in the previous two problems. What minimum PIV rating must the diodes have in this case? Round the answer off to the nearest volt.

✔SOLUTION

The transformer secondary still delivers 234 V RMS AC. In a voltage-doubler circuit, the PIV rating of the diodes should be at least 4.2 times the RMS AC voltage at the transformer secondary. In this case that's 4.2 × 234 PIV, or 983 PIV.

Filtering and Regulation

Most electronic equipment requires something better than the pulsating DC that comes straight out of a rectifier circuit. A *power-supply filter*, also called a *ripple filter*, can minimize or eliminate pulsations in the rectifier output DC.

Capacitors Alone

The simplest power-supply filter consists of one or more large-value capacitors, connected in parallel with the rectifier output, as shown in Fig. 8-5. An *electrolytic capacitor* works well in this role. It's a *polarized* component, meaning that

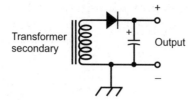

FIGURE 8-5 · A single capacitor can act as a power-supply filter.

we must connect it in a certain direction, just as we would do with a battery or a diode.

Engineers express power-supply filter capacitance values in units called *microfarads*, symbolized µF. Typical electrolytic-capacitor values range from a few microfarads up to several hundred microfarads. The more current that a power supply must deliver, the more capacitance it takes to get effective filtering because large capacitances can hold a charge longer with a given current load than small capacitances can. For a given current demand, the filtering improves as the capacitance increases up to a certain point; beyond that point we get diminishing returns.

Filter capacitors work by "trying" to maintain the DC voltage at its peak level as the output of the rectifier goes through its pulsations. This task is more easily accomplished in the output of a full-wave rectifier (Fig. 8-6A) than in the output of a half-wave rectifier (Fig. 8-6B). With a full-wave rectifier receiving a 60-Hz AC electrical input, the ripple frequency equals 120 Hz. With a half-wave rectifier, the ripple frequency equals 60 Hz.

? Still Struggling

A power-supply filter capacitor recharges twice as often with a full-wave rectifier, as compared with a half-wave rectifier. Therefore, we end up with less ripple at the filter output, for a given amount of capacitance, when we use a full-wave circuit.

Capacitors and Chokes

A superior scheme for "smoothing out" the ripple from a rectifier involves placing a large-value inductor in series with the rectifier output, and a large-value capacitor in parallel with the rectifier output. Engineers express power-supply filter inductance values in units called *henrys* (not "henries"!), symbolized H.

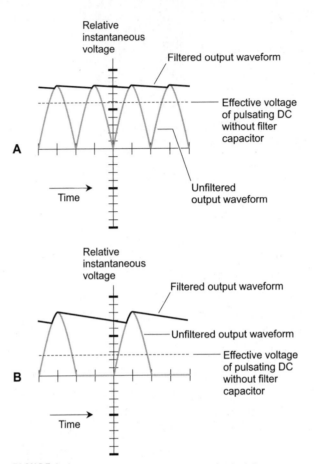

FIGURE 8-6 · Filtering of ripple in the output of a full-wave rectifier (A) and in the output of a half-wave rectifier (B).

The inductor, called a *filter choke*, normally has a value of at least several tenths of a henry, and some of them have values as high as about 10 H. These components are bulky, massive, and expensive, but when we want to build a power supply of the highest possible quality, filter chokes can prove worth the inconvenience and cost.

If we place the capacitor on the rectifier side of the choke, we get a *capacitor-input filter* (Fig. 8-7A). If we place the choke on the rectifier side of the capacitor, we get a *choke-input filter* (Fig. 8-7B). We can use capacitor-input filtering when we don't expect or need a power supply to deliver high current. The output voltage is higher with a capacitor-input circuit than with a choke-input circuit. If we want our power supply to deliver large or variable amounts of current, we're better off using a choke-input filter because the output voltage will remain more stable.

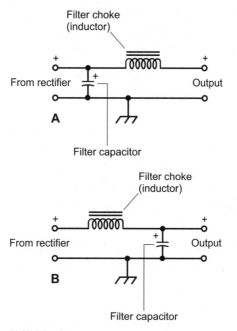

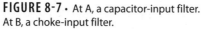

FIGURE 8-7 · At A, a capacitor-input filter. At B, a choke-input filter.

If the output of a DC power supply must have an absolute minimum of ripple, we can connect two or three capacitor/choke pairs in *cascade*, as shown in Fig. 8-8. Each capacitor/choke pair constitutes a *section* of the filter. Multisection filters can consist of either capacitor-input or choke-input sections, but we should never mix the two types in the same filter.

In the example of Fig. 8-8, both capacitor/choke pairs are called *L sections* because of their arrangement in the schematic diagram. If we omit the

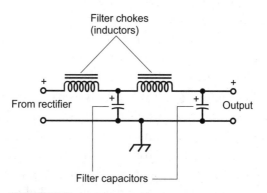

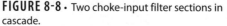

FIGURE 8-8 · Two choke-input filter sections in cascade.

right-hand capacitor, we get a single *T section*. If we replace the left-hand choke with a direct connection, we obtain a single *pi section*. These circuits derive their names from the way they look in schematic diagrams. If we use our imaginations, we can suppose that these component arrangements look like the uppercase English letter L, the uppercase English letter T, and the uppercase Greek letter Π respectively.

TIP *When a power supply operates without any load, or when the load resistance is extremely high, the voltage across the filter capacitors holds steady near the peak rectifier output voltage, not the RMS output voltage. Therefore, the effective voltage across the capacitors exceeds—sometimes greatly—the RMS value of the pulsating voltage that charges them. This dramatic voltage increase explains why, whenever you build a DC power-supply filter, you should use capacitors rated to handle several times the rectifier's effective DC output voltage.*

WARNING! *If you connect an electrolytic capacitor the wrong way around, it won't work as a power-supply ripple filter. In fact, if the backward voltage gets large enough, an electrolytic capacitor can explode like a firecracker! I've seen it happen, and it's dangerous. Always double-check the polarities of all your filter capacitors several times as you build a power supply, and before you apply any power to the system.*

Voltage Regulation

If we connect a specialized semiconductor component called a *Zener diode* in parallel with the output of a power supply along with a resistor in series, the combination of components will limit the output voltage. The Zener diode must have an adequate power rating to prevent it from burning out. The limiting voltage depends on the particular Zener diode used. We can find Zener diodes to fit any reasonable power-supply voltage.

Figure 8-9 is a diagram of a full-wave bridge DC power supply including a capacitor-input filter and a Zener diode for voltage regulation, along with a small-value resistor to limit the current through the diode. The Zener diode's cathode goes to the positive DC output terminal, and its anode goes to the negative terminal. That's opposite from the way we would connect a rectifier diode! We must make sure that we connect the Zener diode in the right direction, or it will fail to regulate the voltage properly, and will most likely burn out the instant that we power-up the circuit.

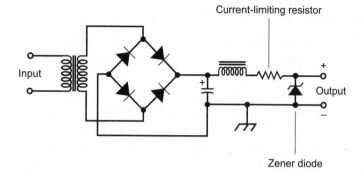

FIGURE 8-9 · A full-wave bridge power supply with a Zener-diode voltage regulator.

A simple Zener-diode voltage regulator will interfere with the operation of a power supply that must deliver a lot of current because the current-limiting resistor "gets in the way." When we need a power supply to deliver high current, we can use a *power transistor* along with a Zener diode to obtain voltage regulation. Figure 8-10 illustrates a voltage-regulator circuit that will work well with high-current power supplies. In this case, we use a so-called *NPN transistor*. The theory and operation of transistors lies beyond the scope of this book. You can learn all about transistors in a comprehensive course, such as *Teach Yourself Electricity and Electronics*. Transistors are also covered in the companion to this book, *Electronics Demystified*.

TIP   *You can find prepackaged voltage regulators in* **integrated-circuit (IC) form.** *Such an IC, sometimes along with some external components, should be installed in the power-supply circuit at the output of the filter. In high-voltage power supplies,* **electron-tube voltage regulators** *are sometimes used.*

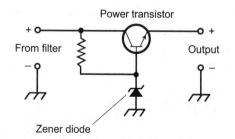
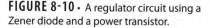

FIGURE 8-10 · A regulator circuit using a Zener diode and a power transistor.

PROBLEM 8-4

Suppose that the standard AC line frequency exceeds 60 Hz. Does the increased frequency make the output of a rectifier circuit easier to filter, or more difficult?

SOLUTION

In theory, an increase in the AC frequency makes the output of a rectifier circuit easier to filter because the capacitors don't have to hold the charge for as long between rectifier-output pulsations as they do at the lower frequency.

PROBLEM 8-5

How can a technician tell when an electrolytic capacitor is wired into a circuit with the polarity correct?

SOLUTION

Most electrolytic capacitors are labeled with either a plus sign or a minus sign, or both. The wires (called *leads*) should be connected in the circuit so that the plus (+) side of the capacitor goes to the part of the circuit with the more positive voltage, or so that the (−) side goes to the part of the circuit with the more negative voltage. If you have an electrolytic capacitor without any polarity markings, don't use it!

PROBLEM 8-6

What will happen if an electrolytic capacitor is connected with the polarity reversed, or to a circuit with a voltage higher than its rated voltage?

SOLUTION

In either of those situations, the component won't provide its rated capacitance. If the applied voltage rises high enough, especially in the reverse direction, an electrolytic capacitor can overheat, rupture, or explode.

Protecting Electronic Equipment

The output of a power supply should never exhibit sudden voltage changes. Such fluctuations can damage equipment or components, or interfere with their proper performance. Voltage must never appear on the external surfaces of a power supply, or on the external surfaces of any equipment connected to it.

Grounding

The best electrical ground for a power supply is the "third wire" ground provided in up-to-date AC utility circuits. The "third hole" (the bottom hole in an AC outlet, shaped like an uppercase English letter D turned on its side) should connect directly to a wire that ultimately terminates in a *ground rod* driven into the earth at the point where the electrical wiring enters the building.

In old buildings, *two-wire AC systems* are common. You can recognize this type of system by the presence of only two slots in the utility outlets. Some of these systems employ reasonable grounding by means of *polarization*, where one slot is longer than the other, and the longer slot goes to electrical ground. But that method is less effective than a *three-wire AC system*, in which the ground connection remains independent of both outlet slots.

TIP *Unfortunately, the presence of a three-wire or polarized outlet system doesn't guarantee that an appliance connected to an outlet will end up well-grounded. If the appliance design is faulty, or if the "third hole" wasn't grounded by the people who installed the electrical system, a power supply can deliver a dangerous voltage to the external surfaces of appliances and electronic devices. This situation can pose an electrocution hazard, and can also hinder the performance of electronic equipment.*

WARNING! *All metal chassis and exposed metal surfaces of AC power supplies should be connected to the grounded wire of a three-wire electrical cord. You should never defeat or cut off the "third prong" of the plug. You should find out whether or not the electrical system in the building was properly installed, so you don't work under the illusion that your system has a good ground when it actually doesn't. If you're in doubt about this issue, have a professional electrician perform a complete inspection of the system.*

Surge Currents

At the instant we turn a power supply on (usually by applying AC to the transformer primary by throwing a switch), a surge of current occurs, even with nothing connected to the supply output. The surge takes place because the filter capacitors need to acquire an initial charge, forcing them to draw a lot of current for a short time. The initial surge current can greatly exceed the normal

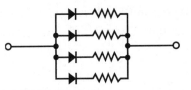

FIGURE 8-11 · Diodes in parallel, with current-equalizing resistors in series with each diode.

operating current. An extreme surge can destroy the rectifier diodes in a poorly designed power supply. We can prevent rectifier-diode failure as a result of current surges in four ways:

1. Use rectifier diodes with current and PIV ratings several times higher than the normal operating level.

2. Make certain that all the rectifier diodes have identical current ratings and identical PIV ratings. Ideally, all the rectifier diodes should bear the same component number.

3. Connect several identical rectifier diodes in parallel wherever the circuit calls for a single diode. *Current-equalizing resistors* of small ohmic value should be connected in series with each diode, as shown in the hypothetical example of Fig. 8-11.

4. Use an automatic switching circuit in the transformer primary. A circuit of this type applies a reduced AC voltage to the transformer for the first second or two after the initial power-up, and then applies the full input voltage.

Transients

The AC on the utility line presents itself as a sine wave with a constant RMS voltage near 117 V or 234 V. But this wave is far from "pure"! If we look at the AC waveform on a high-quality laboratory oscilloscope, we'll occasionally see *voltage spikes*, known as *transients*, that attain positive or negative peak values greatly exceeding the positive or negative peak waveform voltage. Transients result from sudden changes in the load in a utility circuit. A thundershower can produce transients throughout an entire municipality. Unless we take measures to suppress them, transients can destroy the diodes in a power supply. Transients can also interfere with the operation of sensitive electronic equipment, such as computers or microcomputer-controlled appliances.

The simplest way to get rid of common transients involves placing a small-value capacitor of about 0.01 μF, rated for 600 V or more, between each side of the transformer primary and electrical ground, as shown in Fig. 8-12. So-called *disk ceramic* capacitors work well in this application, and cost almost nothing. Commercially made *transient suppressors* are available in most hardware stores and large department stores; they work better than capacitors alone. These devices, often mistakenly called "surge protectors," use specialized semiconductor-based components to prevent sudden voltage spikes from reaching levels where they can cause problems.

TIP *It's a good idea to use transient suppressors with all sensitive electronic devices, including computers, hi-fi stereo systems, and television sets. In the event of a thundershower, the best way to protect such equipment is to physically unplug it from the wall outlets until the storm has passed.*

Fuses

A *fuse* contains a piece of soft wire that melts and falls apart, breaking a circuit if the current exceeds a certain level. You'll normally connect a fuse in series with the transformer primary, as shown in Fig. 8-12, along with transient-suppressing capacitors or a commercially manufactured transient suppressor. A short circuit or overload will burn the fuse out. If a fuse blows out, you must replace it with another fuse having exactly the same current rating. Otherwise, you should expect to encounter one or the other of two problems: frequent and unnecessary fuse burn-out (a mere nuisance) or inadequate equipment protection (a disaster waiting to happen). In a well-designed power supply, the fuse protects the transformer, the diodes, and all the other internal components in case of a short circuit or excessive current demand at the output.

Fuses are available in two types: the *quick-break fuse* and the *slow-blow fuse*. A quick-break fuse contains a straight length of wire or a metal strip. A slow-blow fuse usually has a spring inside along with the wire or strip. You should

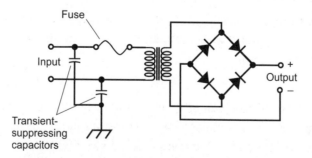
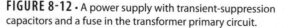

FIGURE 8-12 · A power supply with transient-suppression capacitors and a fuse in the transformer primary circuit.

always replace blown-out fuses with new ones of the same type. Quick-break fuses in slow-blow situations might burn out needlessly (although in some cases they'll work okay). For example, a minor initial current surge, of no cause for concern in normal operation, can burn out a quick-break fuse. Slow-blow fuses in quick-break environments might not provide adequate protection to the equipment, letting excessive current flow for too long before burning out.

Circuit Breakers

A *circuit breaker* performs the same function as a fuse, except that you can easily reset a breaker by turning off the power supply, waiting a moment, and then pressing a button or flipping a switch. Some breakers reset automatically when the equipment has remained powered-down for a certain length of time.

If a fuse or breaker keeps blowing out or tripping, or if it blows or trips immediately after you've replaced or reset it, then trouble exists somewhere in the power supply or in the equipment connected to it. Possible problems include:

- One or more burned-out power-supply diodes
- A bad transformer
- Shorted filter capacitors
- A short circuit in the equipment connected to the supply
- A component connected with the wrong polarity

WARNING! *Never replace a fuse or breaker with a larger-capacity unit to overcome the inconvenience of repeated blowing or tripping. Find the cause of the trouble, and repair the equipment as needed. The "penny in the fuse box" scheme can endanger equipment and personnel, and it increases the risk of fire in the event of a short circuit.*

PROBLEM 8-7

Will a transient suppressor work properly if it's designed for a three-wire electrical system, but the ground wire has been defeated, cut off, or does not lead to a good electrical ground?

SOLUTION

No. In order to function properly, a transient suppressor requires a substantial electrical ground that ultimately leads to a ground rod driven into the earth. The excessive, sudden voltages are shunted away from sensitive equipment only when a current path is provided to allow discharge to ground.

PROBLEM 8-8

What's the difference between a surge and a transient?

✔ SOLUTION

The term *surge* refers to the initial high current drawn by a cheap or poorly designed power supply when you switch it on. The term *transient* refers to a high-voltage spike that can be induced in the AC line by lightning, sparking (known as *arcing*) in utility power transformers, or arcing in equipment connected in the same utility circuit as the supply.

Uninterruptible Power Supply

When we operate an electronic appliance from utility power, a system malfunction or failure can result from a *blackout, brownout, interruption,* or *dip*. We define a blackout as a complete loss of power for an extended period. A brownout represents a significantly reduced voltage for an extended period. An interruption constitutes a complete loss of power for a brief period. A dip is a significantly reduced voltage for a few moments.

To prevent power "glitches" from causing major trouble such as computer data loss, we can use an *uninterruptible power supply* (UPS), such as the one diagrammed in Fig. 8-13. Under normal conditions, the equipment gets its power through the transient suppressor and the regulator. The transient suppressor gets rid of potentially destructive voltage "spikes." The regulator eliminates surges and dips in the utility power. A small current through the rectifier and filter maintains a lead-acid battery in a fully charged state.

If a utility power anomaly or failure occurs, an *interrupt signal* causes the switch to disconnect the equipment from the regulator and connect it to the power inverter, which converts the battery DC output to AC. For the duration of the utility "downtime," the battery discharges; its capacity should be sufficient to last long enough to allow for proper system shutdown. When utility power returns to normal, the switch disconnects the equipment from the battery and reconnects it to the regulator. Then the battery starts to charge up again.

TIP *If power to a computer fails and you have a UPS, save all your work immediately on the hard drive, and also on an external medium, such as a compact disk or flash drive if possible. Then switch the entire system, including the UPS, off until utility power returns, or until you've connected the system to a functioning emergency backup generator.*

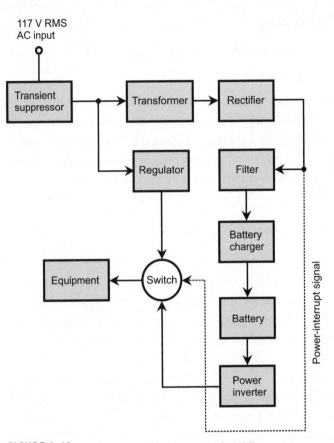

FIGURE 8-13 · An uninterruptible power supply (UPS) prevents system failure in case of utility power interruptions or irregularities.

WARNING! *All electrical power supplies present a potential danger to human life. Some circuits can retain deadly voltages at exposed terminal points (as a result of filter capacitors holding their charge) even after the entire system has been switched off, unplugged, and left unattended for some time. If you have any doubt about your ability to build or work safely with a power supply, leave the task to a professional technician or engineer.*

QUIZ

This is an "open book" quiz. You may refer to the text in this chapter. You'll find the correct answers listed in the back of the book.

1. The ripple frequency at the output of the circuit shown in Fig. 8-14 equals
   A. twice the AC input frequency.
   B. 1.414 times the AC input frequency.
   C. the AC input frequency.
   D. half the AC input frequency.

2. The output of the circuit of Fig. 8-14 has an effective voltage that's
   A. zero.
   B. positive.
   C. negative.
   D. alternating.

3. What appears at the output of the circuit of Fig. 8-15?
   A. A pure DC voltage with positive polarity
   B. A pulsating DC voltage with positive polarity
   C. A pulsating DC voltage with negative polarity
   D. A pure DC voltage with negative polarity

4. What purpose does the pair of components marked X serve in the circuit of Fig. 8-15?
   A. Minimize the ripple in the input
   B. Suppress extreme voltage spikes in the input
   C. Protect the transformer against excessive current
   D. Keep the peak AC input voltage below the PIV rating of the diodes

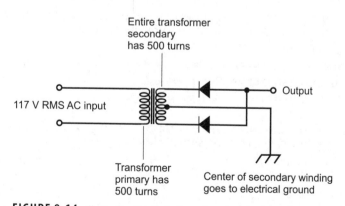

FIGURE 8-14 · Illustration for Quiz Questions 1 and 2.

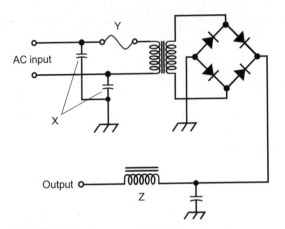

FIGURE 8-15 · Illustration for Quiz Questions 3 through 6.

5. **What function does the component marked Y perform in the circuit of Fig. 8-15?**

   A. Minimize the ripple in the input
   B. Suppress extreme voltage spikes in the input
   C. Protect the transformer against excessive current
   D. Keep the peak AC input voltage below the PIV rating of the diodes

6. **What function does the component marked Z perform in the circuit of Fig. 8-15?**

   A. Prevent excessive output current from flowing
   B. Minimize the ripple in the output
   C. Regulate the output voltage
   D. All of the above

7. **Given a pure sine-wave AC input with no DC component, the effective output voltage from a half-wave rectifier is approximately equal to**

   A. the peak voltage.
   B. half the positive or negative peak voltage.
   C. 71% of the positive or negative peak voltage.
   D. 35% of the positive or negative peak voltage.

8. **If we expect the instantaneous reverse-bias voltage across a diode to reach values of up to 100 PIV, then we should choose a rectifier diode rated at**

   A. 150 PIV or more.
   B. 100 PIV or more.
   C. 71 PIV or more.
   D. 35 PIV or more.

9. If the AC voltage across the secondary winding of the transformer in a full-wave bridge rectifier circuit equals 100 V RMS, then we should choose rectifier diodes rated at

   A. 71 PIV or more.
   B. 100 PIV or more.
   C. 150 PIV or more.
   D. 210 PIV or more.

10. If one of the filter capacitors in a well-engineered full-wave bridge power supply happens to short out, what should we expect?

    A. The filter choke, if any, will burn out.
    B. One or more rectifier diodes will fail.
    C. The transformer will burn out.
    D. The fuse will blow or the breaker will trip.

chapter 9

Wire and Cable

Wire provides a convenient means of getting electricity from one place to another. Wire can also offer physical reinforcement in a variety of situations. Various types of wire serve in the manufacture of lamps, small appliances, resistors, and heating elements. A *cord* or *cable* comprises two or more wires designed to carry currents or signals in specialized applications.

CHAPTER OBJECTIVES

In this chapter, you will

- Contrast the three most common means of expressing relative wire size.
- Learn how a wire's diameter affects its resistance per unit length and its current-carrying capacity.
- Learn to splice and solder wires and cables together.
- Compare different types of electrical cable.
- Compare methods of connecting cables to external devices and to each other.

Wire Conductors

Pure elemental *silver* is the best known electrical conductor at everyday temperatures, followed by *copper* and *aluminum*. In terms of sheer physical strength, *steel* is the strongest conductive material available at moderate cost. Silver, copper, and aluminum, while better electrical conductors than steel, are physically weaker.

Basic Types

The most common type of wire is a rod-shaped length of metal, such as iron, steel, copper, or aluminum. Some wires consist of one conductor, while others have several conductors. A single-conductor wire is called *solid*, and a multi-conductor wire is called *stranded*. Some wires have square or rectangular cross sections, rather than the more common circular cross section.

Wires are sometimes uninsulated (or bare), and sometimes coated with a layer of enamel, rubber, plastic, or other insulating material. Insulated wires work well when we want to avoid short circuits between adjacent conductors. Uninsulated wires are less expensive than insulated wires of the same metal and the same diameter, and are preferable in situations where a short circuit is unlikely or of no consequence.

Stranded wire has some advantages over solid wire. If we have two samples of wire made of the same material and having the same diameter, the stranded wire will be less likely than the solid wire to break when subjected to tension or repeated flexing. The stranded wire will adhere more readily to *solder* than the solid wire will. When installed in long spans, the stranded wire will stretch less than the solid wire will.

On the downside, stranded wire corrodes more quickly than solid wire. When bent, stranded wire doesn't hold its position as well as solid wire does. The effective cross-sectional area of stranded wire is a little less than that of solid wire of the same outside diameter (because of inevitable spaces between strands), so stranded wire has slightly higher resistance per unit length than solid wire of the same outside diameter.

Gauge

Engineers and technicians define wire diameter using a number called the *gauge*. In general, higher gauge numbers correspond with smaller diameters.

The figures apply to the metal only, and do not take into account any insulation or enamel that might coat the metal.

- The *American Wire Gauge* (AWG) is commonly used in the United States. Table 9-1 shows wire diameters for AWG No. 1 through AWG No. 40.

TABLE 9-1 American Wire Gauge (AWG) diameters.

AWG	Diameter in Millimeters	Diameter in Inches
1	7.35	0.289
2	6.54	0.257
3	5.83	0.230
4	5.19	0.204
5	4.62	0.182
6	4.12	0.163
7	3.67	0.144
8	3.26	0.128
9	2.91	0.115
10	2.59	0.102
11	2.31	0.0909
12	2.05	0.0807
13	1.83	0.0720
14	1.63	0.0642
15	1.45	0.0571
16	1.29	0.0508
17	1.15	0.0453
18	1.02	0.0402
19	0.912	0.0359
20	0.812	0.0320
21	0.723	0.0285
22	0.644	0.0254
23	0.573	0.0226
24	0.511	0.0201
25	0.455	0.0179
26	0.405	0.0159

TABLE 9-1 American Wire Gauge (AWG) diameters. *(Continued)*

AWG	Diameter in Millimeters	Diameter in Inches
27	0.361	0.0142
28	0.321	0.0126
29	0.286	0.0113
30	0.255	0.0100
31	0.227	0.00894
32	0.202	0.00795
33	0.180	0.00709
34	0.160	0.00630
35	0.143	0.00563
36	0.127	0.00500
37	0.113	0.00445
38	0.101	0.00398
39	0.090	0.00354
40	0.080	0.00315

- In some countries, the *British Standard Wire Gauge* (NBS SWG) is used. Table 9-2 shows the diameters for NBS SWG designators 1 through 40.

- The *Birmingham Wire Gauge* (BWG) values differ slightly from the American and British Standard values. Table 9-3 shows the diameters for BWG designators 1 through 20.

TABLE 9-2 British Standard Wire Gauge (NBS SWG) diameters.

NBS SWG	Diameter in Millimeters	Diameter in Inches
1	7.62	0.300
2	7.01	0.276
3	6.40	0.252
4	5.89	0.232
5	5.38	0.212
6	4.88	0.192
7	4.47	0.176
8	4.06	0.160

TABLE 9-2 British Standard Wire Gauge (NBS SWG) diameters. *(Continued)*

NBS SWG	Diameter in Millimeters	Diameter in Inches
9	3.66	0.144
10	3.25	0.128
11	2.95	0.116
12	2.64	0.104
13	2.34	0.092
14	2.03	0.080
15	1.83	0.072
16	1.63	0.064
17	1.42	0.056
18	1.22	0.048
19	1.02	0.040
20	0.91	0.036
21	0.81	0.032
22	0.71	0.028
23	0.61	0.024
24	0.56	0.022
25	0.51	0.020
26	0.46	0.018
27	0.42	0.0164
28	0.38	0.0148
29	0.345	0.0136
30	0.315	0.0124
31	0.295	0.0116
32	0.274	0.0108
33	0.254	0.0100
34	0.234	0.0092
35	0.213	0.0084
36	0.193	0.0076
37	0.173	0.0068
38	0.152	0.0060
39	0.132	0.0052
40	0.122	0.0048

TABLE 9-3   Birmingham Wire Gauge (BWG) diameters.

BWG	Diameter in Millimeters	Diameter in Inches
1	7.62	0.300
2	7.21	0.284
3	6.58	0.259
4	6.05	0.238
5	5.59	0.220
6	5.16	0.203
7	4.57	0.180
8	4.19	0.165
9	3.76	0.148
10	3.40	0.134
11	3.05	0.120
12	2.77	0.109
13	2.41	0.095
14	2.11	0.083
15	1.83	0.072
16	1.65	0.064
17	1.47	0.058
18	1.25	0.049
19	1.07	0.042
20	0.889	0.035

Resistance per Unit Length

All wire offers some opposition to the flow of current per unit length. We can express this opposition in terms of a unit called the *micro-ohm per meter*. This unit "works" exactly as its name implies. For example, if a certain type of wire exhibits 15 micro-ohms per meter, then:

- A 1-m length has a resistance of 15 micro-ohms (0.000015 ohm)
- A 1-km length has a resistance of 15,000 micro-ohms (0.015 ohm)
- A 100-km (100,000-m) length has a resistance of 1,500,000 micro-ohms (1.5 ohms)

TABLE 9-4	Resistance per unit length for various gauges of solid copper wire, expressed in micro-ohms per meter and accurate to three significant figures
AWG Wire Size	**Micro-Ohms per Meter**
2	523
4	831
6	1,320
8	2,100
10	3,340
12	5,320
14	8,450
16	13,400
18	21,400
20	34,000
22	54,000
24	85,900
26	137,000
28	217,000
30	345,000

For any particular metal, larger gauges (smaller diameters) of wire have greater resistance per unit length than smaller gauges (larger diameters). In the case of DC at room temperature, we can use Table 9-4 to estimate the resistance per unit length in micro-ohms per meter for even-numbered solid-copper wire diameters from AWG No. 2 through No. 30.

TIP *Resistance per unit length differs from simple resistance. The resistance per unit length of a sample of wire does not depend on how long or short we cut it. The overall resistance of a wire sample, in contrast, increases in direct proportion to the length.*

Current-Carrying Capacity

The ability of a wire to handle DC and utility AC safely is called its *current-carrying capacity* (or simply *carrying capacity*), usually specified in amperes (A). Table 9-5 shows approximate DC and low-frequency RMS AC carrying capacity

TABLE 9-5 Maximum safe continuous DC and utility AC carrying capacity for various American Wire Gauge (AWG) wire sizes, assuming no insulation and an open-air, room-temperature environment.

AWG Wire Size	Current in Amperes
8	73
10	55
12	41
14	32
16	22
18	16
20	11

for even-numbered solid-copper wire sizes from AWG No. 8 through No. 20, in open air at room temperature.

TIP *Wire can intermittently carry larger currents than those shown in Table 9-5, but the danger of softening or melting, with consequent breakage, rises rapidly as the current increases beyond these values. The fire hazard also increases when wire gets too hot.*

CAUTION! *When a span of wire runs alongside electrical components, you should reduce the carrying capacity figures to about half of the values shown in Fig. 9-5. The same precaution applies when wires are bundled into cables, and/or when wires pass near flammable materials or are surrounded by insulation.*

PROBLEM 9-1

How does the DC resistance per unit length of solid wire compare with the wire diameter in linear units, such as millimeters?

SOLUTION

The DC resistance per unit length of solid wire varies directly in proportion to the *reciprocal* (or *inverse*) of the *cross-sectional area*, which equals the surface area of the flat circular region that appears at the end of a solid wire when we cut it straight across. The cross-sectional area varies in direct proportion to the *square* of the diameter. Therefore, the DC resistance per unit length varies in proportion to the *inverse square* of the wire diameter. If we cut the diameter in half, for example, the DC resistance per unit length increases by a factor of 4. If we triple the diameter, the DC resistance per unit length goes down by a factor of 9 (it becomes 1/9 as great as it was before).

PROBLEM 9-2

Solid wire has slightly lower resistance per unit length (it conducts a little better) than stranded wire of the same gauge. Why?

✔ SOLUTION

Imagine a piece of solid wire cut straight across. The cross section is all metal. Now imagine a piece of stranded wire of the same outside diameter (not including any insulation) cut straight across. The cross section isn't all metal; small voids exist because each strand has a disk-shaped cross section, and when we squeeze the disks together, small air gaps inevitably remain. For this reason, the "metallic cross-sectional area" of solid wire exceeds the "metallic cross-sectional area" of stranded wire of the same gauge. Solid wire conducts DC or 60-Hz AC a little better than stranded wire of the same gauge. The resistance per unit length of the solid wire is a little lower than the resistance per unit length of the stranded wire.

Wire Splicing

When we build electrical circuits, we must often splice lengths of wire together. We can use various techniques to accomplish this task. Let's look at the two most common wire-splicing methods, both of which can provide good electrical connections.

Twist Splice

The simplest way to splice two wires involves bringing the exposed ends close together and parallel, and then twisting them over each other several times as shown in Fig. 9-1. We call the resulting connection a *twist splice*.

Twist splicing works with solid wires or stranded wires, as long as they're both of the same type and they both have approximately the same gauge. If the

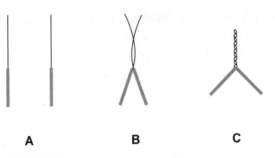

FIGURE 9-1 • Twist splice for wires of the same gauge. Wires are brought parallel (A), looped around each other (B), and then twisted over each other several times (C).

wires differ greatly in diameter, we can wrap the smaller wire around the larger wire (Fig. 9-2), but in that case, we must not let the smaller wire slip off the larger one! We can wrap electrical tape over the connection to help keep it in place, insulate it against the elements, and minimize the risk of short-circuiting with other wires or components.

A twist splice has poor mechanical strength, but it offers convenience for indoor or temporary connections. Twist splices are common inside utility outlets and lamps, where mechanical strength is not an issue. Plastic caps cover the splices to provide insulation, to help hold the wires together, and to streamline the process of "unsplicing" the wires when we want to replace the applicable fixture or component.

Western Union Splice

When a splice must have the maximum possible mechanical strength, the wires are brought together end to end, overlapping about 2 in (5 mm). The wires are

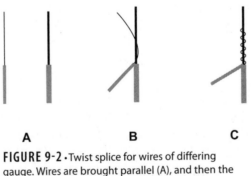

FIGURE 9-2 • Twist splice for wires of differing gauge. Wires are brought parallel (A), and then the smaller wire is looped (B) and twisted (C) around the larger one.

hooked around each other, and then twisted several times, as shown in Fig. 9-3. Technicians call the resulting connection a *Western Union splice*.

For wire spans expected to support weight, you should avoid splices if possible. If splices are necessary, the Western Union method works best. Each end should be twisted around its "mate" 10 to 12 times. Needle-nosed pliers can help to secure the extreme ends. You should finish the connection by removing the protruding wire ends using a diagonal cutter.

Soldering

You can apply *solder* to electrical splices if you need the best possible electrical bond, and if you want the connection to survive for a long time. The *soldering* process can also provide a little extra tensile strength to a Western Union splice by reducing the tendency for the wires to "pull apart" under stress. For large-diameter wires, you should coat both exposed wire ends with solder before making the splice to optimize the electrical bond, a process called *tinning*. For maximum physical strength, both wires should be the same gauge and the same type (both solid, or both stranded).

Before you start to solder a wire splice, give the *soldering gun* or *soldering iron* plenty of time to heat up. Soldering guns generally heat up in a few seconds; soldering irons take a few minutes. Once the instrument has reached its working temperature, hold its tip against each twist in the splice, allowing the wire

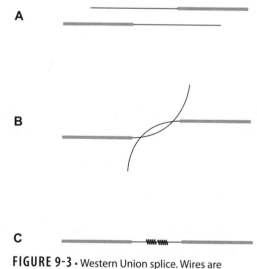

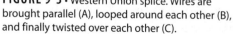

FIGURE 9-3 • Western Union splice. Wires are brought parallel (A), looped around each other (B), and finally twisted over each other (C).

to heat up until the solder flows freely in between the wire turns. Use only enough solder to completely coat or "soak" the connection. Don't use so much solder than any of it drips away from the connection.

If sufficient heat has not been applied to a solder connection, a *cold solder joint* will result. A properly soldered connection has a shiny, clean appearance. A cold joint looks dull or rough. A good solder joint is mechanically immobile ("solid" or "secure"), while a cold one is fragile and can sometimes be broken by merely pulling on it. Many electrical equipment failures occur because of cold solder joints, which can exhibit high resistance and/or intermittent conduction. If you find a cold solder joint in an existing circuit, remove as much of the solder as possible using wire braid called *solder wick* made for that purpose. Then clean the exposed wire surfaces and *resolder* the connection.

TIP   *You'll have to use special* **aluminum solder** *for aluminum wire because aluminum won't adhere to conventional* **soft solder** *intended for copper wire. If you need a particularly strong mechanical bond, you can use* **silver solder.**

Electrical Cable

You'll encounter numerous types of cable intended for the transmission of electrical power or signals over short to moderate spans. Let's look at the most common configurations.

Lamp Cord

The simplest electrical cable, other than a plain single wire, is two-conductor *lamp cord*, which works well with common appliances at low to moderate current levels. Two or three wires are embedded in rubber or plastic insulation (Fig. 9-4A). The individual conductors are stranded to help them resist breakage from repeated flexing. Some household appliance cords have three conductors rather than two. The third wire facilitates electrical grounding.

Multiconductor Cable

When a cable has several wires, they can be individually insulated, bundled together, and enclosed in an insulating jacket (Fig. 9-4B). If the cable must have good flexibility, each wire is stranded. Some cables of this type have dozens of conductors. If only a few conductors exist, they can run parallel to each other in a flat configuration, as shown in Fig. 9-4C. Sometimes, several conductors are molded into a flexible, durable, thin plastic jacket, as shown in Fig. 9-4D, an

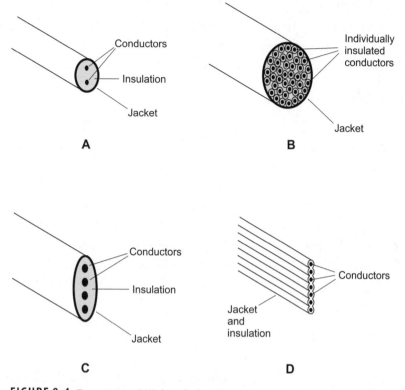

FIGURE 9-4 • Two-wire cord (A), bundled multiconductor cable (B), flat cable (C), and multiconductor ribbon (D).

arrangement known as *multiconductor ribbon*. We'll find this type of cable inside high-tech electronic devices, particularly computers. It's physically sturdy, takes up a minimum of space, and efficiently radiates heat away.

Cable Shielding

The above mentioned cable types are *unshielded*. For DC and utility AC, unshielded cables usually do an acceptable job. But radio signals, video, and high-speed data generate *electromagnetic* (EM) *fields* that can "leak" between conductors within a single cable, and also between a cable and the surrounding environment. In these situations, we need to use cable that has *EM shielding* to prevent the "leakage."

We can shield a wire or cable by wrapping it with a conductive cylinder made of solid metal, metallic braid (usually copper), or metal foil. A layer of insulating material such as polyethylene keeps the shield conductor electrically and physically separated from the main conductor or conductors.

In some multiconductor cables, a single shield surrounds all the wires. In other cables, each wire has its own shield. The entire cable can be surrounded by a copper braid in addition to individual shielding of the wires. *Double-shielded cable* is surrounded by two concentric braids separated by insulation.

Coaxial Cable

Coaxial cable, also called *coax* (pronounced "*co*-ax"), constitutes the transmission medium of choice for *radio-frequency* (RF) AC signal transmission because of its excellent EM shielding characteristics. We can consider any AC whose frequency exceeds a few kilohertz to lie within the so-called *RF spectrum*. We'll see coaxial cable in *community-antenna television* (CATV) networks and in some computer *local area networks* (LANs). Coaxial cable is employed by amateur, marine, Citizens Band, and government radio operators to connect transceivers, transmitters, and receivers to antennas. Coax can also work quite well in high-fidelity sound systems to interconnect components, such as amplifiers, microphones, tuners, and speakers.

In a typical coaxial cable, a cylindrical shield surrounds a single *center conductor*. In some cases, solid or foamed polyethylene insulation, called the *dielectric*, keeps the center conductor at the central axis of the cable (Fig. 9-5A). Other cables have an uninsulated center conductor along with a thin tubular layer of polyethylene just inside the braid (Fig. 9-5B), so that air forms most of the dielectric medium. Some coaxial cables have a solid metal "pipe" or "conduit" surrounding the center conductor. Communications engineers call this type of cable *hard line*. It's available in larger diameters than coaxial cables, and has lower loss per unit length. We'll sometimes encounter hard line in high-power, fixed radio and television transmitting installations.

TIP *In a coaxial cable, the center conductor carries the AC signal. The shield is grounded to keep internal signals from "leaking out." The grounded shield also keeps unwanted external signals from "leaking in."*

Serial and Parallel Cables

We can transmit and receive digital signals, which comprise on/off elements called binary digits (or *bits*), one after another along a single line. This mode constitutes *serial data transmission*. We can obtain higher data speeds by using multiple lines, sending independent bit sequences along each line. In that case, we have *parallel data transmission*. A *serial cable* is designed to carry serial data;

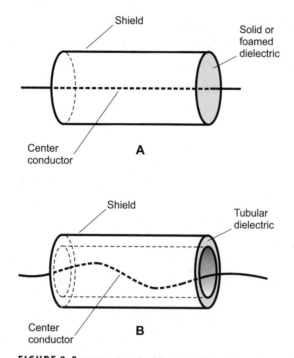

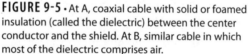

FIGURE 9-5 • At A, coaxial cable with solid or foamed insulation (called the dielectric) between the center conductor and the shield. At B, similar cable in which most of the dielectric comprises air.

a parallel cable is designed to carry parallel data. Not surprisingly, parallel data cables generally have more wire conductors than serial data cables have.

As a general rule, we can make serial cables longer than parallel cables and get good signal-transfer results. The main reason lies in the fact that parallel cables are prone to *crosstalk*, a condition in which supposedly independent bit sequences in multiple adjacent conductors interfere with one another. We can prevent crosstalk by individually shielding each conductor within a parallel cable, but this precaution increases the physical bulk of the cable, and it also increases the cost per unit length.

An example of a serial cable is the coaxial line used to carry television signals in CATV. The cord connecting the main unit and the printer in an older personal computer system is usually a parallel cable. However, in newer computers, a specialized cable called a *universal serial bus* (USB) *cord* connects the main *central processing unit* (CPU) to peripherals, such as printers, scanners, modems, and external drives.

> **TIP** *In recent years, improved serial data transmission technology has rendered parallel cables nearly obsolete in computer systems, except for short runs between circuit boards inside the computer.*

Cable Splicing

When you want to splice cord, multiconductor cable, or ribbon cable, you can use a Western Union splice for each individual conductor. You should wrap all splices individually with electrical tape, and then wrap the entire combination with more electrical tape afterwards. Insulating the individual connections ensures that no two conductors will come into electrical contact with each other inside the splice. Insulating the whole combination keeps the cable from shorting out to external metallic objects, and provides additional protection against corrosion and oxidation.

Two-conductor lamp cord or ribbon cable can be twist-spliced. First, bring the ends of the cables together and pull the conductors outward from the center as shown in Fig. 9-6A. Then twist the corresponding conductor pairs several times around each other as shown in Fig. 9-6B. Solder both connections and then trim the splices to lengths of approximately 1/2 in (1.2 cm). Insulate each twist splice carefully with electrical tape. Then fold the twists back parallel to the cable axis in opposite directions. Finally wrap the entire connection with electrical tape to insulate it and hold the splices in place.

> **TIP** *In order to minimize the risk of a short circuit between conductors in a multi-conductor cable, you can make the splices for each conductor at slightly different*

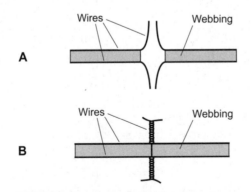

FIGURE 9-6 · Twist splice for two-wire cord or cable. Ends are brought together (A) and the conductors are twisted at right angles (B). The twists are then soldered, trimmed, folded back, and insulated.

points along the cable. In addition, you must ensure that each conductor at the end of the first cable connects to the corresponding conductor at the end of the second cable! This process can get confusing, especially if the color-coding scheme for the wires in the first cable differs from the color-coding scheme for the wires in the second cable. Always double-check your work before insulating the connection.

PROBLEM 9-3

Can coaxial cable effectively carry DC?

SOLUTION

Yes, it can. In some situations, coaxial cable works better as a DC-carrying medium than conventional two-wire lamp cord does. However, coaxial cable costs more per unit length than two-wire lamp cord having conductors of the same carrying capacity. In most situations, lamp cord will suffice as a transmission medium for DC.

PROBLEM 9-4

Under what conditions can we expect coaxial cable to outperform lamp cord for carrying DC or low-frequency AC, such as single-phase utility electricity?

SOLUTION

When we interconnect sensitive electronic devices with lengths of cable, we'll sometimes need to keep EM fields away from the currents flowing in the cable. If the cable conductors are exposed to strong EM fields, the fields can induce additional (and unwanted) current, causing certain types of devices to malfunction or fail. An example is the cable used to connect the speakers of a hi-fi sound system to the amplifier. The use of coaxial cable in place of typical speaker cable (which resembles lamp cord) can eliminate, or at least reduce, the risk that a strong EM field will induce stray AC into the amplifier by means of the speaker cables.

Connectors

A *connector* provides and maintains reliable electrical contact between two different sections of wire or cable. Let's look at some of the most common connectors used in electrical systems.

Appliance Connectors

We're all familiar with the *plugs* attached to the ends of appliance cords. These connectors have two or three prongs, which fit into *outlets* having receptacles in the same configuration. The plugs constitute so-called *male connectors*, and the receptacles into which the plugs fit constitute *female connectors*. (You'll often hear this gender-based terminology when engineers talk about electrical and electronic connectors.)

Appliance plugs and outlets provide temporary electrical connections. They can't offer reliability in long-term applications because the prongs on any plug, and the metallic contacts inside any receptacle, will eventually *oxidize* or *corrode*. A new male plug has shiny metal blades, but an old one has visibly tarnished blades. Similar tarnishing occurs in the "slots" of a female receptacle.

TIP   *Cheap appliance plugs and outlets are intended for indoor use only. You'll need to install special heavy-duty appliance connectors for outdoor use.*

Clip Leads

A *clip lead* is a short length of flexible wire, equipped at one or both ends with a simple, temporary connector. Clip leads don't work well in permanent installations, especially outdoors, because corrosion occurs easily and the connector can slip out of position. The current-carrying capacity is limited. Clip leads are used primarily in DC and low-frequency AC applications.

So-called *alligator clips* often work well for temporary connections in lab tests or experiments. These specialized clip leads require no modification to the circuit under test. The name derives from the serrated edges of the clip sections; the whole connector looks like the mouth of an alligator. Alligator clips come in a wide range of sizes. They can be clamped to electrical terminals, small-diameter pipes or conduits, or lengths of exposed wire.

Banana Connectors

A *banana connector* is a convenient single-lead connector that slips easily in and out of its receptacle. The "business part" of a *banana plug* (male connector) looks something like a banana (or, if you prefer, a cucumber) as shown in Fig. 9-7. *Banana jacks* (female receptacles) are sometimes found inside the *screw terminals* of low-voltage DC power supplies. If you need to change power connections frequently, banana connectors make the task more convenient than the repeated screwing and unscrewing of the power-supply terminals. Banana

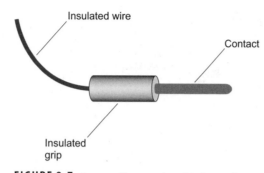

Insulated wire

Contact

Insulated
grip

FIGURE 9-7 · Banana plugs work well in low-voltage
DC applications. The single contact slides into a
cylindrical receptacle (jack).

connectors offer moderately high current-handling capacity at low to moderate
voltages. They, like clip leads and alligator clips, are intended for low-voltage and
short-term indoor use only.

WARNING! *Never use clip leads, alligator clips, or banana connectors in high-*
voltage systems. The exposed conductors present an electrical shock hazard.

Hermaphroditic Connectors

A *hermaphroditic connector* is an electrical plug/jack with two or more contacts, some of them male and some of them female. Usually, hermaphroditic
connectors at opposite ends of a single length of cable appear identical when
you look at them "face-on." However, the pins and holes are arranged in a
special geometry so you can join the two connectors in only the correct way.
This "can't-go-wrong" characteristic makes hermaphroditic connectors ideal
for polarized circuits, such as DC power supplies, and for multiconductor
electrical control cables.

Phone Connectors

Phone plugs and jacks find extensive use in DC and low-frequency AC systems
at low voltages and low current levels. In its conventional form, the male phone
plug (Fig. 9-8A) has a rod-shaped metal *sleeve* that serves as one contact, and a
spear-shaped metal *tip* that serves as the other contact. A ring of hard-plastic
insulation separates the sleeve and the tip. Typical diameters are 1/8 inch (3.175
mm) and 1/4 inch (6.35 mm). The female phone jack (Fig. 9-8B) has contacts
that mate securely with the male plug contacts. The female contacts have built-
in spring action that holds the male connector in place after insertion.

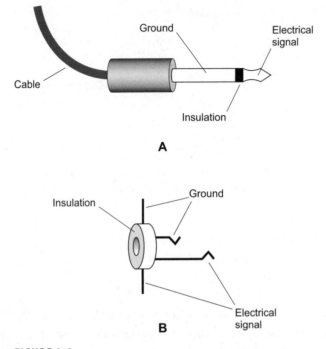

FIGURE 9-8 · At A, a two-conductor phone plug. At B, a two-conductor phone jack.

Engineers originally designed the phone plug and jack for use with two-conductor cables. In recent decades, three-conductor phone plugs and jacks have become common as well. They're used in high-fidelity stereo sound systems and in the audio circuits of multimedia computers and radio receivers. The male plug has a sleeve broken into two parts along with a tip, and the female connector has an extra contact that touches the second sleeve when the plug is inserted into the jack.

TIP   *The term "phone" comes from the original application of phone connectors:* *centralized community* **telephone switchboards** *manned by human operators* *prior to the advent of direct dialing.*

Phono Connectors

Phono plugs and jacks are designed for ease of connection and disconnection of coaxial cable in DC and low-frequency AC systems at low voltages and low current levels. The plug can be pushed on and pulled off. In effect, a phono plug-and-jack combination constitutes a shielded banana plug-and-jack pair.

Phono connectors work in the same situations as phone plugs and jacks do, but they offer better shielding for coaxial-cable connections. Phono plugs and jacks are sometimes called *RCA connectors* after their original designer, the *Radio Corporation of America* (RCA).

TIP   *The term "phono" comes from the earliest design application of phono plugs and jacks: interconnecting the audio components of phonographs ("record players") in the early part of the twentieth century for music recording and reproduction.*

D-Shell Connectors

If a data-transfer cable contains more than three or four conductors, we can connect a *D-shell connector* to either end. These connectors are available in various sizes, depending on the number of wires in the cable. The so-called *ports* on a personal computer, especially the one intended for connecting the *central processing unit* (CPU) to an external *video display*, commonly comprise D-shell connectors. The hardware has a characteristic appearance (Fig. 9-9). The trapezoidal shell forces the user to insert the plug correctly. The female socket has holes into which the pins of the male plug slide. Screws or clips secure the plug once it's in place.

TIP   *Some D-shell connectors have metal outer casings that help keep out dust and moisture, and that also maintain good electrical shielding if needed.*

▢ PROBLEM 9-5

Imagine that you have an appliance with a two-conductor plug on the end of its cord. One of the blades on the plug is slightly wider than the other. Suppose that you live in the United States. The electrical outlets in your house are all

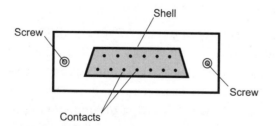

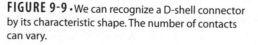

FIGURE 9-9 • We can recognize a D-shell connector by its characteristic shape. The number of contacts can vary.

three-wire types, with two vertical slots and one round or D-shaped hole below the slots. In each outlet, the left-hand slot is slightly wider than the right-hand slot (when viewed so that the round or D-shaped hole appears below the slots). You can plug the two-wire cord into the three-wire outlet, but the plug only goes in one way: the wide blade into the wide slot, and the narrow blade into the narrow slot. What's the reason for the difference in the width of the blades on the plug and the slots in the outlets?

✔SOLUTION

The two-wire plug with unequal-width blades constitutes a *polarized plug*. The wide blade is intended for electrical ground, and the narrow blade is meant to carry single-phase AC to the appliance. In a properly wired home electrical circuit, the functions should match the appliance wiring: the wide slot in each outlet should run to the system ground, and the narrower slot should run to a source of single-phase AC electricity. (The D-shaped hole in a three-wire outlet should also go to the system ground.) A polarized plug and outlet, properly engineered, ensures that the parts of the appliance intended for electrical grounding actually "get grounded," minimizing the risk of electrical shock to people who use the appliance.

PROBLEM 9-6

Imagine that you rent an apartment with wiring hookups for electric laundry machines. You don't own any laundry machines yet, but you've noticed that the electrical outlet for the machines looks different from the regular outlets that you see in other places throughout the house. The laundry outlet has three slots. The two on the top are oriented at a slant, and the one on the bottom is vertical. What type of outlet is this?

✔SOLUTION

Outlets of this sort carry a nominal AC voltage of 234 V RMS, although the actual voltage can vary from about 220 to 240 V RMS. The vertical slot on the bottom goes to the electrical system ground. The slot on the left carries 117 V RMS AC with respect to ground. The slot on the right also carries 117 V RMS AC with respect to ground, but in phase opposition to the AC in the slot on the left. The voltage between the two slanted slots, therefore, equals twice the usual 117 V RMS AC, or 234 V RMS AC.

Why They Don't Cancel

In the situation described in Problem 9-6 and its solution, the two AC voltages, although of equal frequency, equal amplitude, and opposite phase, don't cancel out as you might at first expect. They add together because the output is taken between them—that is, one with respect to the other—rather than by connecting them together and then taking the output with respect to electrical ground. It's like two cars driving toward each other along a two-lane road in opposite directions, one straight northbound and the other straight southbound, each one moving at, say, 11.7 meters per second (m/s). Their average or composite velocity equals zero (we might call northbound travel "positive" and southbound travel "negative"), but the two cars actually move at 11.7×2, or 23.4 m/s with respect to each other.

? Still Struggling

If you were to connect the two 117-V sources in a 234-V utility outlet directly to each other, the voltages would cancel out, but not for long! The sources would "kill each other." This action would short out both 117-V circuits, blowing the fuses or tripping the breakers in both halves of the circuit. (Each 117-V "side" should have its own independent fuse or breaker.) In addition, you'd witness a "flash and bang" at the outlet at the instant you brought the contacts together.

WARNING! *Don't even think about trying the foregoing maneuver as some sort of humorous "mad scientist's experiment." You could end up with a fire in your house wiring (perhaps hidden inside the walls, where you wouldn't find it until it evolved into a real disaster), and the initial point-of-contact spark could burn you or set your clothes on fire.*

PROBLEM 9-7

If a plug becomes corroded or oxidized so that it no longer makes reliable electrical contact with outlets for which it's designed, how can you remedy the problem?

✔ SOLUTION

You can usually tell when the contacts of a plug have corroded or oxidized, because the metal that makes up the prongs appears dark or discolored. You can sometimes remove the oxidation layer by rubbing the contacts with fine-grain sandpaper, emery paper, or steel wool, and then wiping the contacts off with a dry cloth. You should continue the "sanding" process until bright metal shows everywhere on the exposed parts of the contacts. If the contact prongs are too small or too closely spaced for "sanding," then you can use special contact cleaner to remove the oxidation layer. You can find this type of cleaner in a good hardware or electronics store. Sometimes, a solution of common table salt and vinegar will do the job!

TIP   *After cleaning a connector by any means, rinse off the contacts with water and allow them to dry completely before you use the connector.*

QUIZ

This is an "open book" quiz. You may refer to the text in this chapter. You'll find the correct answers listed in the back of the book.

1. Which of the following plug types would we most likely install at the end of a length of coaxial cable?
   A. D-shell
   B. Banana
   C. Phono
   D. Alligator

2. Which of the following four metals conducts electricity the best at room temperature?
   A. Copper
   B. Aluminum
   C. Steel
   D. Silver

3. In a polarized plug intended for use with a standard utility circuit in the United States, the narrower blade should
   A. carry single-phase AC.
   B. connect to ground.
   C. carry three-phase AC.
   D. carry DC.

4. Which of the following connector types offers the simplest option for a temporary DC connection to a length of exposed wire?
   A. D-shell
   B. Banana
   C. Phono
   D. Alligator

5. Which of the following media works best for transmission of RF signals?
   A. Multiconductor ribbon
   B. Coaxial cable
   C. A single wire
   D. Lamp cord

6. Suppose that we have two spans of AWG No. 12 pure copper wire, both measuring exactly 500 m in length. One wire is stranded, and the other wire is solid. We should expect that the stranded wire will
   A. stretch more easily than the solid wire does.
   B. have slightly greater resistance than the solid wire does.
   C. break more easily than the solid wire does.
   D. not adhere to solder as well as the solid wire does.

7. Which of the following characteristics does a twist splice have?

   A. It's convenient for temporary connections.
   B. It offers excellent mechanical strength.
   C. It's difficult to assemble.
   D. It won't work with stranded wires.

8. If a 1000-m span of a certain type and gauge of wire has a resistance of 2.000 ohms, then we should expect that a 1414-m span of the same type and gauge of wire will have a resistance of

   A. 1.414 ohms.
   B. 2.000 ohms.
   C. 2.828 ohms.
   D. 4.000 ohms.

9. If a 1000-m span of a certain type and gauge of wire has 2000 micro-ohms per meter, then we should expect that a 1414-m span of the same type and gauge of wire will exhibit

   A. 1414 micro-ohms per meter.
   B. 2000 micro-ohms per meter.
   C. 2828 micro-ohms per meter.
   D. 4000 micro-ohms per meter.

10. You can often visually recognize a cold solder joint by its

    A. lack of exposed conductors.
    B. excellent mechanical strength.
    C. shiny gold surface.
    D. rough or dull appearance.

Test: Part II

Do not refer to the text when taking this test. You may draw diagrams or use a calculator if necessary. A good score is at least 38 correct. You'll find the answers listed in the back of the book. Have a friend check your score the first time, so you won't memorize the answers if you want to take the test again.

1. Suppose that a load constitutes a pure resistance of 50 ohms. We measure a certain constant RMS AC voltage across it. Then we increase the RMS voltage to four times its previous value. Assuming that the resistance does not change, the amount of power that the load dissipates

   A. remains the same.
   B. doubles.
   C. increases by a factor of 4.
   D. increases by a factor of 8.
   E. increases by a factor of 16.

2. Which of the following characteristics is an advantage of a power supply with a full-wave bridge rectifier over a power supply with a half-wave rectifier?

   A. It's easier to get rid of the output ripple.
   B. It places less strain on the transformer.
   C. It places less strain on the diode(s).
   D. The voltage drops less when we demand high current output.
   E. All of the above

3. Figure Test II-1 shows a waveform as it might appear on a laboratory oscilloscope. Each vertical division represents 1.00 V, and each horizontal division represents 100 ms. What's the positive peak voltage?

   A. +0.30 V pk+
   B. +0.75 V pk+
   C. +1.50 V pk+
   D. +3.00 V pk+
   E. +7.50 V pk+

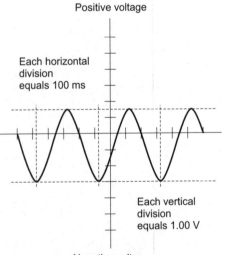

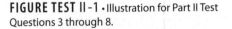

FIGURE TEST II-1 • Illustration for Part II Test Questions 3 through 8.

4. What's the negative peak voltage of the wave shown in Fig. Test II-1?
   A. −0.60 V pk–
   B. −1.50 V pk–
   C. −3.00 V pk–
   D. −6.00 V pk–
   E. −15.0 V pk–

5. What's the peak-to-peak voltage of the wave shown in Fig. Test II-1?
   A. 4.50 V pk-pk
   B. 6.00 V pk-pk
   C. 2.25 V pk-pk
   D. 5.00 V pk-pk
   E. 2.00 V pk-pk

6. Assuming that the waveform in Fig. Test II-1 precisely follows the shape of a sinusoid, what's its average voltage?
   A. +1.50 V avg
   B. +0.75 V avg
   C. 0.00 V avg
   D. −0.75 V avg
   E. −1.50 V avg

7. What's the period of the waveform shown in Fig. Test II-1?
   A. 40 ms
   B. 80 ms
   C. 120 ms
   D. 200 ms
   E. 400 ms

8. What's the frequency of the waveform shown in Fig. Test II-1?
   A. 1.25 Hz
   B. 2.5 Hz
   C. 4.0 Hz
   D. 8.0 Hz
   E. 25 Hz

9. Voltage-multiplier power supplies generally perform well only when
   A. the load never draws much current.
   B. a lot of output ripple can be tolerated.
   C. choke-input filters are included.
   D. the required output voltage is low.
   E. transformers won't work.

10. If we apply a steadily increasing DC bias to a semiconductor diode with the anode at a *more negative* voltage than the cathode, we'll eventually reach a potential difference at which the diode begins to conduct. We call this threshold the

    A. reverse-bias cutoff.

    B. peak forward voltage.

    C. avalanche voltage.

    D. forward breakover voltage.

    E. tolerance limit.

11. If we apply a steadily increasing DC bias to a semiconductor diode with the anode at a *more positive* voltage than the cathode, we'll eventually reach a potential difference at which the diode begins to conduct. We call this threshold the

    A. reverse-bias cutoff.

    B. peak forward voltage.

    C. avalanche voltage.

    D. forward breakover voltage.

    E. tolerance limit.

12. Figure Test II-2 illustrates an AC transformer with a tap in the center of the secondary winding. The primary winding has exactly 80 turns. The entire secondary has exactly 40 turns; the tap divides it into two sections of exactly 20 turns each. We connect a constant, pure sine-wave AC source of 100 V RMS to the primary winding. What AC voltage *X* should we expect to observe across the "upper" portion of the secondary winding?

    A. 20 V RMS

    B. 25 V RMS

    C. 35 V RMS

    D. 50 V RMS

    E. 71 V RMS

13. In the scenario of Fig. Test II-2, what AC voltage *Y* should we expect to observe across the entire secondary winding?

    A. 25 V RMS

    B. 35 V RMS

    C. 50 V RMS

    D. 71 V RMS

    E. 100 V RMS

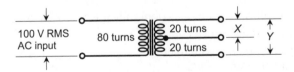

FIGURE TEST II-2 • Illustration for Part II Test Questions 12 and 13.

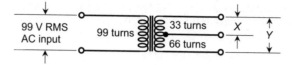

FIGURE TEST II-3 · Illustration for Part II Test Questions 14 and 15.

14. Figure Test II-3 illustrates an AC transformer with a tap in the secondary winding. The primary winding has exactly 99 turns. The tap divides the secondary into an "upper" section with 33 turns and a "lower" section with 66 turns. We connect a constant, pure sine-wave AC source of 99 V RMS to the primary winding. What AC voltage X should we expect to observe across the "upper" portion of the secondary winding?

   A. 11 V RMS
   B. 33 V RMS
   C. 50 V RMS
   D. 66 V RMS
   E. 99 V RMS

15. In the scenario of Fig. Test II-3, what AC voltage Y should we expect to observe across the entire secondary winding?

   A. 198 V RMS
   B. 99 V RMS
   C. 66 V RMS
   D. 50 V RMS
   E. 33 V RMS

16. The resistance per unit length of solid, pure aluminum wire varies in direct proportion to the

   A. diameter.
   B. radius.
   C. cross-sectional area.
   D. temperature.
   E. None of the above

17. If a generator produces 4.50 kW of electrical output power when we provide it with 5000 W of mechanical driving power, then its efficiency equals approximately

   A. 45.0%.
   B. 50.0%.
   C. 90.0%.
   D. 95.0%.
   E. 97.5%.

18. We can place a large-value capacitor across the *DC output* terminals of a full-wave center-tap rectifier circuit in order to

   A. protect the diodes against transients.
   B. protect the load from excessive current.
   C. maintain a constant output voltage.
   D. reduce the amount of ripple in the output.
   E. protect the transformer in case of a short circuit at the output.

19. We can place small-value capacitors between the *AC input* terminals of a full-wave center-tap rectifier circuit and electrical ground in order to

   A. protect the diodes against transients.
   B. protect the load from excessive current.
   C. maintain a constant output voltage.
   D. reduce the amount of ripple in the output.
   E. protect the transformer in case of a short circuit at the output.

20. Suppose that two perfect AC sine waves, both having the same frequency, occur in phase coincidence. Neither wave contains any superimposed DC component. One wave has 44.0 V pk-pk, and the other wave has 11.0 V pk-pk. The composite wave has

   A. 11.0 V pk-pk.
   B. 27.5 V pk-pk.
   C. 33.0 V pk-pk.
   D. 44.0 V pk-pk.
   E. 55.0 V pk-pk.

21. In the scenario of Question 20, the approximate positive peak voltage of the composite wave is

   A. +11.0 V pk+.
   B. +27.5 V pk+.
   C. +33.0 V pk+.
   D. +44.0 V pk+.
   E. +55.0 V pk+.

22. Consider an AC utility outlet that provides 234-V RMS single-phase AC electricity at a frequency of 60 Hz with no DC component superimposed. What's the positive peak voltage?

   A. +82.7 V pk+
   B. +165 V pk+
   C. +331 V pk+
   D. +468 V pk+
   E. +662 V pk+

23. Suppose that we change the frequency of the AC wave described in Question 22 to 240 Hz, but we leave all the other wave characteristics the same. The peak-to-peak voltage

   A. doesn't change.
   B. increases by a factor of 2.

   C. increases by a factor of 4.

   D. becomes 1/2 as great.

   E. becomes 1/4 as great.

24. **Figure Test II-4 shows a schematic diagram of a**

   A. full-wave bridge rectifier circuit.

   B. voltage-doubler rectifier circuit.

   C. half-wave rectifier circuit.

   D. quarter-wave rectifier circuit.

   E. power inverter.

25. **If the AC at the input of Fig. Test II-4 has a frequency of 60 Hz, then the output has a ripple frequency of**

   A. 30 Hz.

   B. 60 Hz.

   C. 120 Hz.

   D. 240 Hz.

   E. zero, because it's pure DC.

26. **An AC phase shift of 60° represents**

   A. 1/36 of a cycle.

   B. 1/24 of a cycle.

   C. 1/12 of a cycle.

   D. 1/6 of a cycle.

   E. 1/4 of a cycle.

27. **Which of the following metals is physically strongest for use as electrical wire at moderate expense?**

   A. Steel

   B. Copper

   C. Aluminum

   D. Silver

   E. All of the above metals have equal physical strength for use as electrical wire at moderate cost.

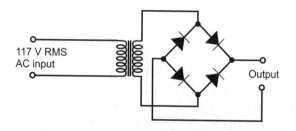

FIGURE TEST II-4 • Illustration for Part II Test Questions 24 and 25.

28. If you want to join two lengths of wire end-to-end so that the resulting span has the maximum possible physical strength, you would want to use

A. a twist splice with no solder.
B. a Western Union splice with no solder.
C. banana connectors.
D. a Western Union splice with silver solder.
E. a twist splice with conventional solder.

29. What's the frequency of an AC wave whose peak-to-peak amplitude equals 2.00 mV?

A. 50.0 Hz
B. 500 Hz
C. 5.00 kHz
D. 50.0 kHz
E. We need more information to calculate it.

30. If a 100-m span of a certain type and gauge wire has a resistance of 3.00 ohms, then we should expect that a 400-m span of the exact same type and gauge of wire will have a resistance of

A. 0.750 ohms.
B. 1.50 ohms.
C. 3.00 ohms.
D. 6.00 ohms.
E. 12.0 ohms.

31. If a 900-m span of a certain type and gauge wire exhibits 36.0 micro-ohms per meter, then we should expect that a 300-m span of the exact same type and gauge of wire will have

A. 4.00 micro-ohms per meter.
B. 12.0 micro-ohms per meter.
C. 108 micro-ohms per meter.
D. 324 micro-ohms per meter.
E. None of the above

32. The frequency of an AC square wave varies inversely in proportion to its

A. decay time.
B. RMS amplitude.
C. average amplitude.
D. period.
E. rise time.

33. Figure Test II-5 shows a schematic diagram of a

A. full-wave bridge rectifier circuit.
B. voltage-doubler rectifier circuit.
C. half-wave rectifier circuit.
D. quarter-wave rectifier circuit.
E. power inverter.

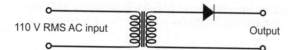

FIGURE TEST II-5 · Illustration for Part II Test Questions 33 and 34.

34. **If the AC at the input of Fig. Test II-5 has a frequency of 50 Hz, then the output has a ripple frequency of**

A. 25 Hz.

B. 50 Hz.

C. 100 Hz.

D. 200 Hz.

E. zero, because it's pure DC.

35. **Imagine two spans of pure copper wire, both 100 m long and having a size of AWG No. 10. One conductor is solid and the other conductor is stranded. How do the DC resistances of the spans compare?**

A. The solid-wire span has much lower DC resistance than the stranded-wire span.

B. The solid-wire span has slightly lower DC resistance than the stranded-wire span.

C. The two spans have equal DC resistances.

D. The solid-wire span has slightly higher DC resistance than the stranded-wire span.

E. The solid-wire span has much higher DC resistance than the stranded-wire span.

36. **Suppose that two perfect AC sine waves, both having the same frequency, occur in phase opposition. Neither wave contains any superimposed DC component. One wave has 50 V pk-pk, and the other wave has 20 V pk-pk. The composite wave therefore has**

A. 15 V pk-pk.

B. 20 V pk-pk.

C. 30 V pk-pk.

D. 35 V pk-pk.

E. 70 V pk-pk.

37. **In the scenario of Question 36, the approximate negative peak voltage of the composite wave is**

A. −15 V pk−.

B. −20 V pk−.

C. −30 V pk−.

D. −35 V pk−.

E. −70 V pk−.

38. **Which of the following cord or cable types would work best in a community antenna television (CATV) system?**

A. Lamp cord

B. Coaxial cable

C. Multiconductor ribbon

D. Single wire

E. Any of the above; they'd all work equally well.

39. Which of the following connector types would we most likely install at the end of a single length of wire to facilitate a temporary connection for low-voltage DC?

   A. D-shell
   B. RCA
   C. Alligator clip
   D. Phone
   E. Hermaphroditic

40. If two waves appear in phase coincidence but offset by a full cycle, we can say, in strict technical terms, that they differ by a phase angle of

   A. 45°.
   B. 60°.
   C. 90°.
   D. 180°.
   E. 360°.

41. Figure Test II-6 is a schematic diagram of a power supply that produces pure DC output when we apply an AC input at 117 V RMS. What does component X do?

   A. It protects the rectifier diodes against damage from transients.
   B. It helps to regulate the output voltage.
   C. It prevents excessive current from flowing through component Z.
   D. It maximizes the output ripple frequency.
   E. It helps to eliminate ripple in the output.

42. In the circuit of Fig. Test II-6, what does component Y do?

   A. It protects the rectifier diodes against damage from transients.
   B. It ensures a constant ripple frequency at the output.
   C. It prevents excessive current from flowing through component Z.
   D. It maximizes the output ripple frequency.
   E. It helps to eliminate ripple in the output.

43. In the circuit of Fig. Test II-6, what does component Z do?

   A. It helps to regulate the output voltage.
   B. It eliminates ripple in the output.

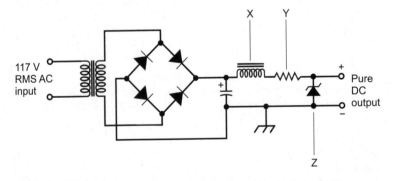

FIGURE TEST II-6 • Illustration for Part II Test Questions 41 through 43.

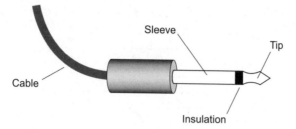

FIGURE TEST II-7 · Illustration for Part II Test Question 46.

    C. It ensures that the output polarity remains correct.
    D. It protects the rectifier diodes against damage in case of a short circuit at the output.
    E. All of the above

44. **What's the frequency of an AC wave whose period equals 667 ns?**
    A. 10.0 MHz
    B. 5.00 MHz
    C. 3.00 MHz
    D. 1.50 MHz
    E. We need more information to calculate it.

45. **Suppose that you have a pair of electrical devices with cables intended for direct connection to each other. Each cable has a connector with several contacts, some male and some female. You're probably looking at**
    A. RCA connectors.
    B. phono connectors.
    C. hermaphroditic connectors.
    D. phone connectors.
    E. banana connectors.

46. **We would most likely use the connector shown in Fig. Test II-7**
    A. to connect a headset to the output of an audio amplifier.
    B. at the output of a high-voltage DC power supply.
    C. at the input of a power supply that operates from 117 V RMS AC.
    D. to connect an external video display to the main unit of a computer.
    E. to connect an appliance to a utility outlet.

47. **A transformer is designed to step 117 V RMS utility AC up to 468 V RMS AC for use in the power supply for a high-power tube-type hi-fi amplifier. The transformer primary has 100 turns. How many turns does the secondary have?**
    A. 200
    B. 400
    C. 800
    D. 1600
    E. We need more information to answer this question.

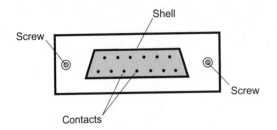

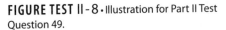

FIGURE TEST II-8 • Illustration for Part II Test
Question 49.

48. Suppose that you live in the United States. The electrical outlets in your house are all three-wire types, with two vertical slots and one round or D-shaped hole below the slots. In each outlet, the left-hand slot is slightly wider than the right-hand slot (when viewed so that the round or D-shaped hole appears below the slots). Assuming that the electricians who installed the house wiring did their jobs correctly, which of the following statements is true?

   A. The larger vertical slot and the D-shaped hole both run to electrical ground, while the smaller vertical slot carries three-phase 117 V RMS AC.
   B. The D-shaped hole runs to electrical ground, the larger vertical slot carries 117 V RMS AC, and the smaller vertical slot carries 117 V AC in the opposite phase.
   C. The two vertical slots and the D-shaped hole each carry 117 V RMS AC, with each current separated by 120° of phase with respect to the other two.
   D. The two vertical slots and the D-shaped hole each carry 234 V RMS AC, with each current separated by 120° of phase with respect to the other two.
   E. The larger vertical slot and the D-shaped hole both run to electrical ground, while the smaller vertical slot carries single-phase 117 V RMS AC.

49. We would most likely use the connector shown in Fig. Test II-8

   A. to connect a headset to the output of an audio amplifier.
   B. at the output of a high-voltage DC power supply.
   C. at the input of a power supply that operates from 117 V RMS AC.
   D. to connect an external video display to the main unit of a computer.
   E. to connect an appliance to a utility outlet.

50. What's the period of an AC wave whose frequency equals 8.00 MHz?

   A. 125 ns
   B. 1.25 µs
   C. 12.5 µs
   D. 125 µs
   E. 1.25 ms

Part III

Magnetism

10

What Is Magnetism?

Magnetic effects relate closely with electrical effects. *Magnetism* exists whenever electric *charge carriers*, such as electrons, move with respect to other objects, or with respect to an observer's frame of reference.

CHAPTER OBJECTIVES

In this chapter, you will

- Learn what causes magnets to attract and repel certain metals and other magnets.
- Define magnetic polarity and see how it relates to electric current.
- Discover how engineers define total magnetic field quantity.
- Learn how current and displacement relate to magnetic field strength.
- See how the earth's magnetic field makes a hiker's or mariner's compass work.
- Learn how the sun's behavior can affect the earth's magnetic field.

Magnetic Force

As children, most of us discovered that certain metal objects, called *magnets*, "stick" to iron and steel. Iron, nickel, and alloys containing either or both of these elements are known as *ferromagnetic materials*. Magnets attract objects made from these metals or alloys. We don't observe the *magnetic force* between magnets and *nonferromagnetic* metals, such as copper or aluminum, unless an electric current flows through the metal. Electrical insulators, such as wood, plastic, and glass, don't interact with magnets under normal conditions.

Cause and Strength

When we bring a magnet near a piece of ferromagnetic material, the atoms in the material line up to some extent, temporarily magnetizing the material and producing a magnetic force between the atoms of the ferromagnetic substance and those in the magnet.

If we bring a magnet near another magnet, we observe more force than we do when we bring a magnet near a sample of a ferromagnetic metal that hasn't been magnetized beforehand. The force between two magnets can be either *repulsive* (the magnets push away from each other) or *attractive* (the magnets pull towards each other), depending on how we orient the magnets relative to each other.

Every magnet has poles called *north* (N) and *south* (S), just as a battery has poles called positive (+) and negative (−). As with electrically charged objects, pairs of similar magnetic poles (N-N or S-S) produce repulsive force, and pairs of opposite poles (S-N or N-S) produce attractive force. The force, whether repulsive or attractive, increases as we bring the magnets closer together.

TIP *Some magnets have strength so great that no human can pull them apart if they get stuck together with poles S-N or N-S, and no person can push a pair of them all the way together against their mutual repulsive force with poles N-N or S-S. In contrast, some magnets have so little strength that we can't notice the force except with the help of scientific instruments.*

Charge Carriers in Motion

Whenever the atoms in a sample of ferromagnetic material are aligned, a *magnetic field* exists around that object. A magnetic field can also result from the motion of electric charge carriers through any conductive medium, such as a wire, or through *free space* (a vacuum). The charge carriers are usually electrons. However, moving protons, atomic nuclei, or charged atoms called *ions* can also produce magnetic fields.

The magnetic field around a *permanent magnet* arises from the same fundamental cause as the field around a wire that carries an electric current. The responsible factor in either case is the motion of electrically charged particles. In a wire, the electrons move among the atomic nuclei in more or less the same direction. In a permanent magnet, the movement of electrons around the nuclei occurs in such a manner that an *effective current* results from their motion within the individual atoms.

Lines of Flux

Physicists consider magnetic fields to comprise *flux lines*, also called *lines of flux*. We can quantify magnetic field intensity according to the number of flux lines passing through a flat region having a certain cross sectional area, usually 1 square centimeter (1 cm^2) or 1 square meter (1 m^2). The flux lines are abstractions, not real objects, but they help scientists to define and express magnetic *field geometry* and *field strength*.

Have you ever placed iron filings on a horizontal sheet of paper and then brought a magnet near the filings, holding the magnet underneath the paper to keep the filings from flying up and sticking to the magnet? The filings arrange themselves in a pattern that shows the shape of the magnetic field in the vicinity of the magnet. A *bar magnet* has a field whose lines of flux have a characteristic, symmetrical pattern (Fig. 10-1). In this illustration, the flux lines appear as dashed curves. The flux lines converge or diverge at the ends of the magnet where the poles exist.

Another interesting magnetism experiment involves passing a current-carrying wire at a right angle through a horizontally oriented piece of paper, and then scattering iron filings on the paper. The filings group themselves in concentric circles around the point where the wire passes through the paper, as shown in Fig. 10-2. This experiment demonstrates that, in any plane passing through a straight, current-carrying conductor at a 90° angle, the magnetic lines of flux form circles centered on the wire.

Polarity

A magnetic field possesses a specific direction (or orientation) at any point in space near a current-carrying wire or a permanent magnet. The flux lines run parallel with the direction of the field. Physicists consider a magnetic field to

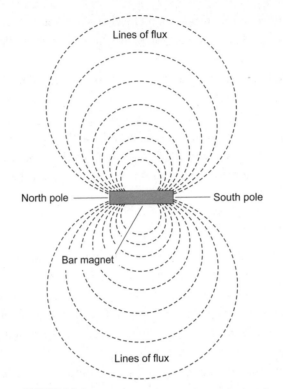

FIGURE 10-1 · Magnetic flux (dashed curves) around a bar magnet (rectangle).

originate at a north pole and to terminate at a south pole. Usually, magnetic lines of flux diverge from north poles and converge toward south poles. Certain exceptions to this rule exist, however, such as in the case of flux surrounding a straight wire.

Interestingly, the north and south poles of a permanent magnet don't coincide with the north and south magnetic poles of the earth (the *geomagnetic poles*). In fact, they directly oppose. The *north geomagnetic pole* is actually a south magnetic pole because it attracts the north poles of permanent magnets. The *south geomagnetic pole* is a north magnetic pole because it attracts the south poles of permanent magnets.

TIP *You can easily locate the magnetic poles in a bar magnet. You'll have a harder time locating or defining the poles near a straight, current-carrying wire because the magnetic field revolves in an endless, circular path. Nevertheless, you can define the direction of flux around a straight wire, using a rule called Ampere's Law.*

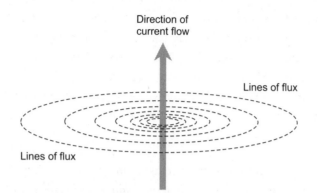

FIGURE 10-2 · Magnetic flux (dashed curves) produced by charge carriers traveling in a straight line.

Ampere's Law

Remember, physicists define theoretical current, also called conventional current, as flowing from the positive electric pole to the negative electric pole (plus to minus). Imagine that the conventional current in Fig. 10-3 flows straight out of the page toward you, at a right angle to the plane containing the paper. According to Ampere's Law, the magnetic flux revolves counterclockwise in this situation, in planes parallel to the plane containing the paper.

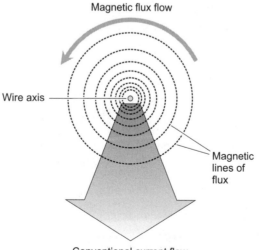

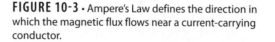

FIGURE 10-3 · Ampere's Law defines the direction in which the magnetic flux flows near a current-carrying conductor.

Ampere's Law is sometimes called the *right-hand rule*. If you hold your right hand with the thumb pointing out straight and the fingers curled, and then point your thumb in the direction of the conventional current flow in a straight wire, your fingers curl in the direction of the magnetic flux flow. Similarly, if you orient your right hand so that your fingers curl in the direction of the magnetic flux flow and then straighten out your thumb, your thumb points in the direction of the conventional current.

TIP *If you want to determine the direction of magnetic flux flow relative to the motion of electrons (from minus to plus), use your left hand instead of your right hand. Then execute the same maneuvers as you would do with your right hand, but "in mirror-image form."*

Monopoles and Dipoles

A charged electric particle, hovering in space and not moving, constitutes an *electric monopole* (meaning an isolated positive or negative pole, without an opposite pole anywhere nearby). The *electric flux lines* near an electric monopole are never closed curves. In fact, they form radial lines, as shown in Fig. 10-4. When a positive electric pole exists near a negative pole, we have an *electric dipole*. The electric flux lines near an electric dipole are always closed curves.

It's tempting to think that the magnetic field around a current-carrying wire arises from a monopole, or that no magnetic poles exist at all because the concentric circles apparently don't originate or terminate anywhere. They don't converge toward any point or diverge from any point. However, we can indirectly define the poles.

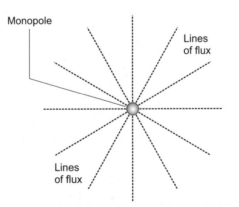

FIGURE 10-4 • Lines of flux near a monopole. Illustration for Problems 10-1 and 10-2.

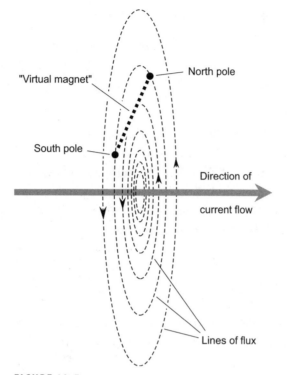

FIGURE 10-5 · A "virtual magnet" near a current-carrying wire.

Imagine two distinct points on one of the flux circles in a perpendicular plane intersecting the wire. We can imagine that a *magnetic dipole*, or pair of opposite magnetic poles, is implied by the flux flow along the *minor arc* (the part of the flux circle that goes less than halfway around) from one point to the other. Let's call this pair of poles a "virtual magnet" because it has no physical form; it's nothing more than a geometric line segment connecting two points in space (Fig. 10-5). In this drawing, the conventional current runs from left to right.

Still Struggling

The lines of flux in the vicinity of a magnetic dipole always connect the two poles. Some flux lines are straight in a local sense, but in the larger sense they always form closed curves. The greatest magnetic field strength around a permanent magnet prevails near the poles, where the flux lines converge or diverge. Around a current-carrying wire, the greatest magnetic field strength exists near the wire.

PROBLEM 10-1

Suppose that we discover a way to create a magnetic monopole. Imagine a point that forms a defined magnetic north or south pole, but no opposite pole exists anywhere nearby. What would the lines of flux around such a point look like?

✔ SOLUTION

The lines of flux would be straight, and they would radiate from (or converge toward) the point in three dimensions. Figure 10-4 is a two-dimensional diagram of this situation. The magnetic flux appears as dashed lines. If a magnetic monopole were placed on the page, the page were oriented horizontally, and iron filings were scattered on the page, the filings would orient themselves along the dashed lines.

PROBLEM 10-2

Suppose that the magnetic monopole shown in Fig. 10-4 were a north pole. Would the lines radiate outward or converge inward? What if the monopole were a south pole?

✔ SOLUTION

If the monopole were a north pole, the flux lines would radiate away from it. If the monopole were a south pole, the flux lines would converge toward it.

Magnetic Field Quantity and Strength

We can express the overall extent or quantity of a magnetic field in units called *webers*, symbolized Wb. We can use a smaller unit, the *maxwell* (Mx), to express the quantity of a weak magnetic field. One weber equals 100,000,000 maxwells, and one maxwell equals 0.00000001 weber. Using power-of-10 notation, we can write these facts as

$$1 \text{ Wb} = 10^8 \text{ Mx}$$

and

$$1 \text{ Mx} = 10^{-8} \text{ Wb}$$

TIP   *To convert an expression of magnetic field quantity from webers to maxwells, multiply by 10^8. To convert an expression of magnetic field quantity from maxwells to webers, multiply by 10^{-8} or divide by 10^8.*

The Tesla and the Gauss

You'll sometimes hear or read about the "strength" of a magnet in terms of webers or maxwells. But more often, you'll hear or read about units called *teslas* (T) or *gauss* (G). These units express the concentration, or intensity, of the magnetic field within a certain cross section, a quantity known as *flux density*. As a variable in equations, we denote flux density as B. A flux density of one tesla equals one weber per meter squared. A flux density of one gauss equals one maxwell per centimeter squared. We can write these facts in equation form as

$$1\ T = 1\ Wb/m^2$$

and

$$1\ G = 1\ Mx/cm^2$$

As things work out, one tesla equals 10,000 gauss and one gauss equals 0.0001 tesla. In power-of-10 notation, these facts are stated as

$$1\ T = 10^4\ G$$

and

$$1\ G = 10^{-4}\ T$$

TIP *To convert an expression of magnetic flux density from teslas to gauss (not gausses!), multiply by 10^4. To convert an expression of magnetic flux density from gauss to teslas, multiply by 10^{-4} or divide by 10^4.*

?

Still Struggling

Does the distinction between magnetic quantity and flux density confuse you? Imagine a lamp inside an enclosed chamber. Suppose that the lamp emits 20 W of visible-light power. If you surround the lamp completely by the walls, floor, and ceiling of the chamber, then 20 W of visible light strike the interior surface, regardless of the chamber's size. But this fact doesn't give us a very useful notion of the lamp's actual brilliance. A 20-W lamp emits plenty of light for a small closet, but it's nowhere near adequate to illuminate a huge auditorium. If you want to know how well the lamp will illuminate the chamber, you must know the light's intensity in terms of watts *per unit area*. When you say that a lamp gives off a certain number of visible-light watts overall, it's like saying that a magnet has an overall field quantity of so-many webers or maxwells. When you say that a lamp causes a certain number of visible-light watts to fall on a given surface area, it's analogous to saying that a magnetic field has a flux density of so-many teslas or gauss within a well-defined surface region.

The Ampere-Turn and the Gilbert

In some situations, we'll need magnetic units other than the weber, maxwell, tesla, or gauss. The *ampere-turn* (At) expresses the intensity of a phenomenon called *magnetomotive force*, which we can imagine as the magnetic equivalent of electromotive force.

A length of wire, bent into a single-turn loop and carrying 1 A of current, produces 1 At of magnetomotive force. If we coil the same wire into a loop having 50 turns and the current stays the same, the resulting magnetomotive force increases by a factor of 50, and therefore, becomes 50 At. If we reduce the current in the 50-turn loop to 1/50 A or 20 mA, the magnetomotive force decreases by a factor of 50 and goes back down to 1 At.

Engineers sometimes employ a unit called the *gilbert* (Gb) to express magnetomotive force. The gilbert is slightly smaller than the ampere-turn. One gilbert equals approximately 0.7958 ampere-turns, and one ampere-turn equals approximately 1.257 gilberts. Mathematically,

$$1 \text{ Gb} = 0.7958 \text{ At}$$

and

$$1 \text{ At} = 1.257 \text{ Gb}$$

Magnetomotive force is usually associated with *electromagnets*, which derive their magnetic "power" from applied electric currents. In mathematical equations, we can symbolize magnetomotive force as F_m. Electromagnets find widespread application in industrial systems, medical devices, engineering development, and scientific research. We'll see how they work in Chapter 12.

TIP *To approximate the magnetomotive force in gilberts when you know the number of ampere-turns, multiply by 1.257. To approximate the magnetomotive force in ampere-turns when you know the number of gilberts, multiply by 0.7958.*

PROBLEM 10-3

Imagine that we drive a current of 30.0 mA through a coil having 230 turns. What's the magnetomotive force F_m in ampere-turns? In gilberts?

SOLUTION

To get the magnetomotive force in ampere-turns, we multiply the current by the number of turns. First, we must convert the current to amperes: 30.0 mA = 0.0300 A. Then we can calculate

$$F_m = 0.0300 \times 230$$
$$= 6.90 \text{ At}$$

To convert this figure to gilberts, we multiply by 1.257, getting

$$F_m = 6.90 \times 1.257$$
$$= 8.67 \text{ Gb}$$

PROBLEM 10-4

What happens to the magnetomotive force if we double the diameter of the coil described in Problem 10-3, but we don't change the current or the number of turns? What happens if we wrap the wire around a ferromagnetic core, again leaving the current and the number of turns constant?

SOLUTION

The magnetomotive force doesn't change. It depends only on the current and the number of turns, not on the coil diameter (or its shape, or its length, or any of its other physical dimensions). Magnetomotive force doesn't depend on what we place inside the coil, either. We could wrap the wire around a powdered-iron core and still end up with the same amount of magnetomotive force as we get if the coil exists in free space, as long as we don't change the current or the number of turns.

Flux Density versus Current

In a straight wire carrying a steady, direct current and surrounded by free space (air or a vacuum), the flux density is greatest near the wire, and diminishes with increasing distance from the wire. We can use a formula to express flux density as a function of distance from the wire. Like all formulas in physics, it's perfectly accurate only under idealized circumstances.

Imagine an "idealized wire" that's infinitely thin, arbitrarily long, absolutely straight, and surrounded by free space. Suppose that the wire carries a current of I amperes. Consider a point P at a distance r (in meters) from the wire, as measured along the shortest possible path (within a plane perpendicular to the wire), as shown in Fig. 10-6. Let B_t represent the flux density (in teslas) at point P. Then

$$B_t = 2 \times 10^{-7}\, I/r$$

As long as the thickness of the wire remains tiny compared with the distance r from it, and as long as the wire remains straight in the vicinity of the point P at which we measure the flux density, this formula will provide us with a good indication of what happens around real-world electrical conductors carrying current. The coefficient of 2 in the above formula represents a mathematically exact value, good to as many significant figures as we need.

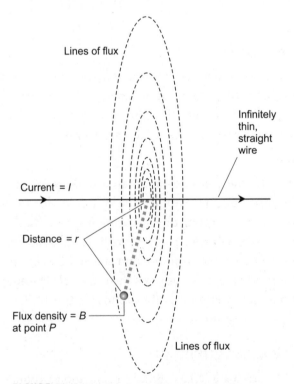

FIGURE 10-6 • The flux density B at a point P near a current-carrying wire varies with the current I in the wire, and also with the distance r from the wire.

PROBLEM 10-5

What's the flux density B_t in teslas at a distance of 200 mm from a straight, thin wire carrying 400 mA DC?

SOLUTION

Before we start calculating, we must convert the distance to meters and the current to amperes, getting $r = 0.200$ m and $I = 0.400$ A. Then we can plug these numbers into the formula and calculate to obtain

$$B_t = 2 \times 10^{-7}\, I/r$$
$$= 2 \times 10^{-7}\, (0.400/0.200)$$
$$= 4.00 \times 10^{-7}\, \text{T}$$

PROBLEM 10-6

In the above scenario, what's the flux density B_g in gauss at point P?

SOLUTION

To solve this problem, we must convert from teslas to gauss, so we multiply the answer from the previous problem by 10^4 to get

$$B_g = 4.00 \times 10^{-7} \times 10^4$$
$$= 4.00 \times 10^{-3}\, \text{G}$$
$$= 0.00400\, \text{G}$$

Geomagnetism

The earth has a core made up largely of superheated iron, some of it in the liquid state. As the earth rotates, this molten iron flows in complicated convection patterns, giving rise to the so-called *geomagnetic field* that surrounds our planet and extends millions of kilometers into space.

Earth's Magnetic Poles and Axis

The geomagnetic field has poles, just as a bar magnet does. The *geomagnetic poles* lie some distance away from the *geographic poles*, which constitute the points on the surface through which the earth's *rotational axis* cuts. The *north geomagnetic pole* (which is actually a south magnetic pole, as previously discussed) is located in extreme northern Canada. The *south geomagnetic pole* is in the ocean near the coast of Antarctica. The *geomagnetic axis* is, therefore, tilted relative to the axis on which the earth rotates. In addition, the geomagnetic axis doesn't pass exactly through the center of the earth as the geographic axis does.

Solar Wind

Charged subatomic particles from the sun constantly stream outward through the solar system. Astronomers and physicists call this phenomenon the *solar wind*. It distorts the geomagnetic field, in effect "blowing" the magnetic lines of flux out of symmetry. On the side of the earth facing the sun, the geomagnetic lines of flux are compressed. On the side of the earth opposite the sun, the geomagnetic lines of flux are dilated. Figure 10-7 illustrates this distortion. The

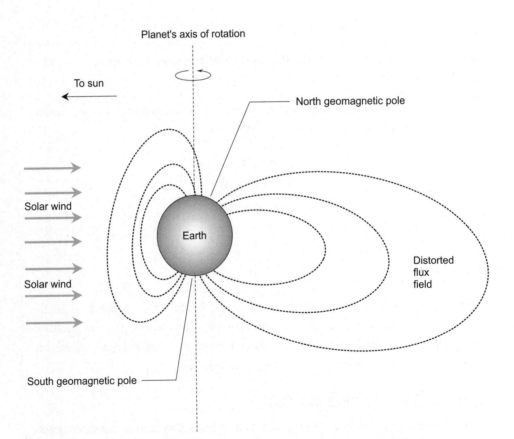

FIGURE 10-7 · The solar wind "blows" the geomagnetic field into an asymmetrical shape.

same effect occurs with the magnetic fields around some other planets in the solar system, particularly Jupiter, which generates a magnetic field many times stronger than the one around the earth.

TIP *At and near the earth's surface, the lines of flux are nearly symmetrical with respect to the geomagnetic poles. As the distance from the earth increases, the extent of the distortion increases. As the earth rotates on its geographic axis, the geomagnetic axis wobbles, causing the geomagnetic field to do a "twist-and-turn dance" through space in the direction facing away from the sun.*

The Magnetic Compass

Magnetized iron-containing rocks called *lodestones* orient themselves in a generally north-south direction when allowed to turn freely. In ancient times, people attributed this phenomenon to a "mystery force" in the atmosphere. The effect took place throughout the civilized world. Seafarers and land explorers found it useful. Even nowadays, a *magnetic compass* can serve as a navigation aid for hikers, mariners, and others who travel far from familiar landmarks. It will continue to work when more sophisticated navigational devices fail.

The geomagnetic field interacts with the field around a magnetized compass needle, producing force on the needle. The force works not only in a horizontal plane (parallel to the earth's surface), but also vertically in most locations. The vertical component of the force equals zero at the *geomagnetic equator*, a line running around the globe equidistant from both geomagnetic poles. As the *geomagnetic latitude* increases, either towards the north or the south geomagnetic pole, the magnetic force pulls up and down on the compass needle more and more. The vertical angle, in degrees relative to the horizon, of the magnetic flux lines at any particular location is called the *inclination* of the geomagnetic field at that location. The compass needle "wants" to align itself parallel to the geomagnetic lines of flux in three-dimensional space, so one end of the needle dips down toward the compass face while the other end tilts up toward the glass.

A magnetic compass rarely points exactly toward geographic north because the north geomagnetic pole doesn't coincide with the north geographic pole. We define the *declination* as the difference, in degrees, between the horizontal-plane (or *azimuth*) orientation of a compass needle and true geographic north. The declination varies, depending on your location.

TIP   *If you try to use a compass at or near the north geomagnetic pole, the needle won't orient itself in any particular direction, so the compass will fail to work. No one ever noticed this phenomenon (or had a problem with it) until compass-carrying explorers ventured into the North American Arctic. The same thing happens at or near the south geomagnetic pole.*

Charge Carriers in Space

The particles of the solar wind carry positive electric charge. Because of their charge and motion, these particles each produce a small effective current. The combined effect of all the moving particles amounts to a vast conventional current flowing outward from the sun. The current produced by each moving, charged particle generates a magnetic field. When the magnetic fields produced by the particles interact with the geomagnetic field, the particles accelerate toward the geomagnetic poles. By "accelerate," we mean that the particles not only change speed, but they change direction as well. The particle acceleration gives rise to an *electromagnetic* (EM) *field*. We'll learn about EM fields in the next chapter.

If an eruption called a *solar flare* takes place on the sun, our "parent star" ejects more solar wind than normal. When the charged particles approach the geomagnetic poles, the "solar hurricane" becomes a powerful cosmic radio-wave transmitter that can wreak havoc on human activities on the surface. The accelerating charged particles cause *ionization* (electrical charging) of the atoms in the earth's upper atmosphere, spawning a so-called *geomagnetic storm* that can affect long-distance radio communication and broadcasting. If the storm grows intense enough, it can disrupt wire-based and cable-based communications and, in the worst cases, the electric utility grid.

TIP   *If you live far from the equator and away from a big city, you've seen the* aurora borealis *(northern lights) or* aurora australis *(southern lights) at night during geomagnetic storms. During an aurora event, ionized atoms in the upper atmosphere* fluoresce *(glow) because of the extra energy they derive from the solar wind.*

PROBLEM 10-7

Draw a diagram showing how the earth's rotational axis, the geomagnetic poles, and geomagnetic lines of flux would appear if no solar wind existed.

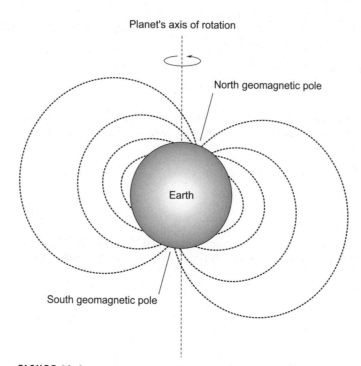

FIGURE 10-8 · Illustration for Problem 10-7.

![checkmark icon] **SOLUTION**

See Fig. 10-8. Note that the earth has, in effect, a huge bar magnet running through it at a slant and slightly off center. For practical purposes (such as navigation using a magnetic compass), we can imagine that the geomagnetic poles lie on the earth's surface.

QUIZ

This is an "open book" quiz. You may refer to the text in this chapter. You'll find the correct answers listed in the back of the book.

1. In the scenario of Fig. 10-9, suppose that the wire carries 10 A of current, and point *P* lies 100 mm away from the wire axis. What's the flux density in gauss at point *P*?

    A. 0.10 G
    B. 0.20 G
    C. 0.40 G
    D. 0.80 G

2. In the scenario of Fig. 10-9, suppose that the wire carries 10 A of current, and point *Q* lies 50 mm away from the wire axis. What's the flux density in gauss at point *Q*?

    A. 0.10 G
    B. 0.20 G

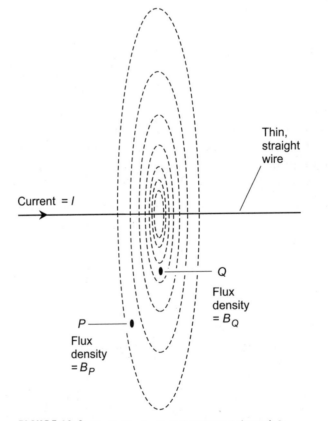

FIGURE 10-9 · Illustration for Quiz Questions 1 through 3.

C. 0.40 G

D. 0.80 G

3. **In the scenario of Fig. 10-9, suppose that the wire carries 20 A of current, and point Q lies 50 mm away from the wire axis. What's the flux density in gauss at point Q?**

A. 0.10 G

B. 0.20 G

C. 0.40 G

D. 0.80 G

4. **Which of the following scenarios can produce a repulsive magnetic force?**

A. The north pole of a permanent magnet brought near a piece of iron

B. The south pole of a permanent magnet brought near a piece of glass

C. The north pole of a permanent magnet brought near the south pole of another permanent magnet

D. None of the above

5. **The magnetomotive force in the vicinity of a wire coil depends on the**

A. number of turns in the coil.

B. radius of the coil.

C. type of material around which we wrap the wire.

D. end-to-end length of the coil.

6. **Suppose that steady DC flows through a straight wire, such that the electrons move directly toward us. We see the magnetic flux as flowing**

A. in straight lines, going radially outward from the wire.

B. in circles, going counterclockwise.

C. in straight lines, running parallel to the wire and going away from us.

D. in circles, going clockwise.

7. **At a specific location on the earth's surface, the vertical angle at which the geo-magnetic flux lines slant, relative to the horizon, tells us the**

A. inclination.

B. locations of the magnetic poles.

C. flux density.

D. effective current.

8. **In the vicinity of a magnetic dipole, such as we would observe in a bar magnet, the flux lines invariably take the form of**

A. straight lines.

B. open arcs.

C. closed curves.

D. perfect circles.

9. Imagine that we drive a current of 100 mA through a coil having 100 turns. What's the magnetomotive force F_m in gilberts, accurate to three significant figures?

A. 12.6 Gb

B. 7.96 Gb

C. 10.0 Gb

D. We must know the coil radius to calculate it.

10. The south geomagnetic pole

A. lies exactly at the south geographic pole.

B. constitutes a magnetic north pole.

C. attracts a magnetic north pole.

D. does not interact with a magnetic compass.

Electromagnetic Effects

An electric current can produce a magnetic field, and a magnetic field can induce electric currents in moving conductors. Under some circumstances, magnetic fields can generate electric fields and vice-versa. Let's look at some of the physical effects common to both electricity and magnetism.

CHAPTER OBJECTIVES

In this chapter, you will

- See how a current-carrying wire can affect the behavior of a magnetic compass.
- Build a simple galvanometer.
- Learn how electromagnetic cathode-ray tubes and oscilloscopes work.
- Discover how electromagnetic fields arise.
- Learn what "ELF radiation" is, and what it can do.

Electromagnetic Deflection

When you place a magnetic compass near a wire carrying DC, the compass needle deflects. The extent of deflection depends on the distance between the compass and the wire, and also on how much current the wire carries. Scientists call this phenomenon *galvanism* or the *electromagnetic effect*.

The Galvanometer

When experimenters first discovered galvanism, they tested various arrangements to obtain the greatest possible current-detecting sensitivity. When the scientists wrapped the wire in a coil around the compass, with the axis of the coil aligned east-west (and the plane of the coil, therefore, aligned north-south), they got a *galvanometer* that could detect the presence of small currents.

A galvanometer takes advantage of the electromagnetic effect. The device resembles an ammeter, but the needle of the galvanometer rests in the center position instead of at the left-hand end of a graduated scale. Current flowing in one direction results in deflection of the needle to the right; current in the other direction causes the needle to move to the left.

You can build a galvanometer by using a compass and some insulated wire. The best wire for this purpose is solid, soft-drawn, insulated copper *bell wire*. You can find this type of wire in any good hardware store. I recommend AWG No. 20 or No. 22. Wind the wire 8-1/2 times around the compass along the north-south line on its scale, as shown in Fig. 11-1. Set the compass down on a nonferromagnetic, horizontal surface, such as a wooden table. Turn the compass

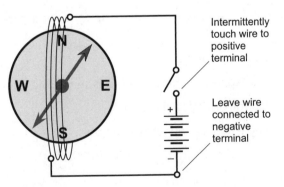

FIGURE 11-1 · You can construct a galvanometer by wrapping insulated wire several times around a magnetic compass.

so that the wire coil and the needle line up with each other. Then connect a flashlight cell to the coil for a second or two, and watch the needle deflect. Reverse the polarity of the cell and repeat the test. The needle will deflect in the opposite direction. Don't leave the cell connected to the coil for more than a couple of seconds at a time.

The compass needle deflects because the magnetic field produced by the current in the coil is oriented at a right angle to the earth's magnetic field. When the coil carries DC, the compass "sees" the magnetic poles as combinations of the geomagnetic poles and the poles of the field set up by the coil. As the current increases, the coil's magnetic field grows stronger with respect to the geomagnetic field, and the compass needle, therefore, deflects to a greater extent.

TIP *In my book* Electricity Experiments You Can Do at Home *(McGraw-Hill, 2010), I describe the construction and testing of a simple galvanometer, along with some experiments that demonstrate the principles of magnetism.*

? Still Struggling

For a specific number of coil turns and a particular compass, the extent of the needle deflection (in angular degrees from geomagnetic north) varies according to the current that the coil carries, but not in direct proportion. The deflection-versus-current function is *nonlinear*, meaning that the relation between needle deflection and coil current shows up as a curve, not as a straight line, when plotted on a coordinate system. Figure 11-2 is a graph of the deflection-versus-current function for a galvanometer that I built in my home lab by wrapping 8-1/2 turns of wire around a hiker's compass. The small open circles represent experimental results.

PROBLEM 11-1

Suppose that a galvanometer coil carries 20 mA of DC, causing the compass needle to deflect 20° to the west of geomagnetic north. What will the needle do if we reverse the direction in which the current flows, but the current remains at 20 mA?

✔ SOLUTION

The compass needle will deflect 20° to the east of geomagnetic north.

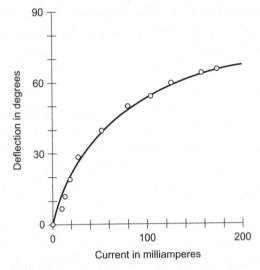

FIGURE 11-2 · In a galvanometer, the relationship between coil current and needle deflection appears as a curve when graphed.

PROBLEM 11-2

What will a galvanometer do if we connect a source of 60-Hz household AC to the coil of a galvanometer, with a large-value resistor in series to ensure that the current doesn't heat up the wire too much, or produce a magnetic field of excessive strength?

SOLUTION

The compass needle will point toward geomagnetic north, just as it would do if no current flowed through the coil. We might notice a slight vibration in the needle as it attempts to move to the west of north, then to the east, and then to the west again, over and over, as the magnetic force reverses every 1/120 of a second.

The Electromagnetic Cathode-Ray Tube

An *electromagnetic cathode-ray tube* (or *electromagnetic CRT*) provides an example of electromagnetic deflection that's quite different in practice from the magnetic compass. In the electromagnetic CRT, a so-called *electron gun* emits a thin beam of electrons inside an evacuated chamber. Donut-shaped electrodes called *anodes*, each carrying a positive DC voltage, accelerate (speed up) and focus the electron beam. The motion of the electrons produces current that gives rise to a magnetic field, just as the flow of current through a wire produces a magnetic field around the wire. After the electrons have been accelerated by the anodes, the beam passes through magnetic fields produced by pairs of *deflection coils* situated at right angles to each other. The beam changes direction as it passes the coils because of the interaction between the magnetic fields from the coils and the magnetic field produced by the moving electrons. The electrons finally strike a screen having an inner surface coated with phosphor. The impinging electrons cause the phosphor to glow in a well-defined spot.

Unless the current through the deflection coils moves the electron beam back and forth, up and down, or in some combination of these directions, we'll only see a fixed, bright spot in the center of the CRT screen. Signals applied to the deflection coils make meaningful displays possible. Figure 11-3 is a simplified functional diagram of an electromagnetic CRT. To minimize clutter, only one pair of deflection coils is shown. As the current through the coils varies in intensity, the magnetic field between the coils fluctuates in strength, and the

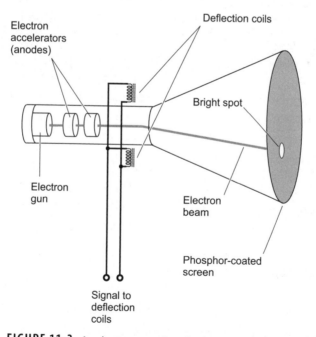

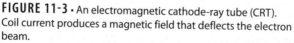

FIGURE 11-3 · An electromagnetic cathode-ray tube (CRT). Coil current produces a magnetic field that deflects the electron beam.

electron beam deflects to a greater or lesser extent. If we supply AC to the coils, then the polarity of the magnetic field alternates, and the electron beam shifts direction alternately back and forth. In the situation of Fig. 11-3, the deflection coils are positioned above and below the electron beam. Coil current, therefore, causes the beam to deflect either toward us or away from us, depending on the polarity of the signal that we apply. We must connect the coils to the signal terminals in identical directions, so that the magnetic fields from the coils work with each other (*aiding*) rather than against each other (*bucking*).

The Cathode-Ray Oscilloscope

The electromagnetic CRT forms the "heart" of a lab instrument called a *cathode-ray oscilloscope*. The *horizontal deflection coils* receive a sawtooth wave, which causes the beam to sweep across the screen at a controllable, constant speed from left to right, return almost instantaneously, and then sweep across the screen from left to right again, over and over. The *vertical deflection coils* receive the signal that we want to analyze. This signal causes the beam to deflect up and down.

We can use an oscilloscope to analyze complicated electrical waves. We can adjust the *sweep frequency* (the number of times per second that the beam makes a complete left-to-right trip, or sweep, across the screen), allowing us to examine waveforms over a wide range of frequencies. We can adjust the *vertical sensitivity* by switching various resistors in series with the vertical deflection coils. The sweep-frequency and vertical-sensitivity controls make it possible to fit almost any waveform, regardless of frequency or amplitude (within reason), neatly on the screen.

TIP   *In a well-designed oscilloscope, specialized circuits synchronize the sweep frequency at some whole-number fraction of the input-signal frequency, so that the waveform stays fixed on the screen instead of constantly racing or jerking toward the left or right. Engineers call this synchronization process* **triggering.**

Electromagnetic Fields

Charged particles, such as electrons and protons, are surrounded by *electric fields*. Magnetic poles or moving charged particles produce *magnetic fields*. When the fields grow strong enough, their effects can extend over considerable distances through space. When electric or magnetic fields vary in intensity, an *electromagnetic* (EM) *field* results.

Static Fields

We've all observed the attractive force between opposite poles of magnets, and the repulsive force between identical poles. Similar effects occur with electrically charged objects.

Electrical forces operate only over short distances under laboratory conditions because such fields rapidly weaken as the distance between poles increases. In theory, however, an electric field extends into space indefinitely unless something blocks it, such as a grounded enclosure made of conducting wire mesh or conducting sheet metal.

A permanent magnet, or a wire carrying a constant electric current, produces a magnetic field that rapidly weakens as we get farther from its origin. In theory, a magnetic field extends into space forever unless something blocks it, such as an enclosure made of ferromagnetic sheet metal.

The existence of a constant voltage difference between two nearby objects, or a constant current in a wire, produces a static electric field (also called an *electrostatic field*) or a static magnetic field (which we might call a *magnetostatic field*, but we'll rarely if ever hear an engineer use that term).

? Still Struggling

A constant voltage or a constant direct current, all by itself, doesn't produce an EM field. In order for an EM field to exist near a charged particle or a current-carrying wire, the voltage or current intensity must fluctuate or alternate.

Fluctuating Fields

A varying voltage between two nearby objects, or a variable current in a conducting medium, gives rise to an EM field. We can think of the EM field as a "dance" between a fluctuating magnetic field and a fluctuating electric field. The changing magnetic field gives rise to a changing electric field, which in turn gives rise to another changing magnetic field. The process keeps repeating so that the electric and magnetic fields "leapfrog" through space at the speed of light. In a vacuum, this speed, denoted by the lowercase italic letter c, equals approximately 299,792 km/s (186,282 mi/s). Engineers sometimes round it off to 3.00×10^5 km/s or 1.86×10^5 mi/s.

An EM field can travel, or *propagate*, tremendous distances. It dies off less rapidly than either a static electric field or a static magnetic field. The electric and magnetic lines of flux in an EM field run perpendicular to each other at every point in space. The *direction of propagation* of an EM field goes perpendicular to both the electric and magnetic flux lines (Fig. 11-4).

In order for an EM field to exist, electric charge carriers must not only move, but they must *accelerate* (speed up) or *decelerate* (slow down). The most common method of creating this sort of situation is the introduction of an alternating current (AC) in an electrical conductor. An EM field can also result from the deflection of charged-particle beams by electric or magnetic fields.

All wires that carry AC, even household utility wires, produce EM fields. When charged particles from the sun follow curved tracks through space near the geomagnetic poles, EM fields result. These fields occur over a wide range of frequencies, and they can attain stupendous overall magnitude. As we learned in Chapter 10, these EM fields can have widespread effects on human activities, occasionally disrupting communications systems and utility power grids.

Frequency and Wavelength

As the frequency of an EM field in free space (a vacuum) increases, the *wavelength* (physical distance between wave crests) decreases, as shown in Fig. 11-5. The frequency and the wavelength are inversely proportional. Their product equals the speed of light. Figure 11-5A shows a hypothetical EM wave. At B, the frequency has doubled compared with the scenario at A, so the wavelength

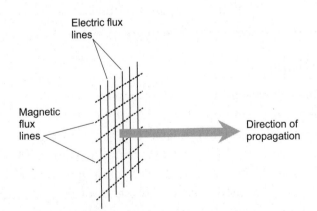

FIGURE 11-4 • Propagation of an EM field. At every point in space, the EM field travels in a direction perpendicular to both the electric lines of flux and the magnetic lines of flux.

FIGURE 11-5 · At A, an EM wave in space. At B, the wavelength becomes half as great when we double the frequency. At C, the wavelength becomes 1/4 as great when we quadruple the frequency.

drops by half. At C, the frequency has quadrupled with respect to the situation at A, and the waves are only 1/4 as long.

At 1 kHz, the wavelength of an EM field in free space equals approximately 300 km. At 1 MHz, the wavelength is about 300 m. At 1 GHz, the wavelength is about 300 mm. At 1000 GHz, an EM signal has a wavelength of 0.3 mm, a distance so small that we would need a magnifying glass to resolve it if we could make a "printout" of it. The frequency of an EM wave can greatly exceed 1000 GHz, and the corresponding wavelength can get far shorter than 0.3 mm.

Scientists have observed EM waves with wavelengths measuring as small as 0.00001 *Ångström* (10^{-5} Å). An Ångström represents 10^{-10} m. Some physicists use the Ångström to express or define exceptionally short EM wavelengths. We'd need a microscope of great magnifying power to see an object with a length of 1 Å. Another unit, increasingly favored by scientists and engineers in recent years, is the *nanometer* (nm), which represents a thousand-millionth of a meter. Therefore

$$1 \text{ nm} = 10^{-9} \text{ m}$$

$$= 10 \text{ Å}$$

and

$$1 \text{ Å} = 10^{-10} \text{ m}$$

$$= 0.1 \text{ nm}$$

When we want to denote wavelength as a variable in equations, we write a lowercase, italic Greek letter lambda (λ). We denote frequency as a lowercase, italic English letter f. In free space, the wavelength λ (in meters) of an EM field relates to the frequency f (in hertz) according to the formula

$$\lambda = c/f$$
$$= 3.00 \times 10^8/f$$

We can use this same formula for λ in millimeters and f in kilohertz, for λ in micrometers and f in megahertz, and for λ in nanometers and f in gigahertz. We recall that a millimeter (1 mm) equals 10^{-3} m, a micrometer (1 μm) equals 10^{-6} m, and a nanometer (1 nm) equals 10^{-9} m.

The formula for frequency f (in hertz) as a function of the wavelength λ (in meters) for an EM field in free space is

$$f = c/\lambda$$
$$= 3.00 \times 10^8/\lambda$$

As in the preceding case, this formula will work for f in kilohertz and λ in millimeters, for f in megahertz and λ in micrometers, and for f in gigahertz and λ in nanometers.

Many Forms

The discovery of EM fields ultimately led to all the wireless communications systems we know today. *Radio waves* were the earliest useful manifestation of EM radiation, but these waves do not represent the only form that an EM field can take! As the frequency increases above that of conventional radio, we encounter *microwaves*. As the frequency keeps going up, we come to *infrared* (IR) radiation, also imprecisely called "heat rays." After that comes *visible light*, then *ultraviolet* (UV) radiation, then *X rays*, and finally *gamma rays*. Sound (acoustic) waves, ocean waves, and shock waves in solid objects do not occur as EM fields, but as mechanical disturbances in matter.

TIP *Electromagnetic fields can exist at frequencies far below those of conventional radio signals. In theory, an EM wave can go through one complete cycle every hour, day, year, thousand years, or million years. Some astronomers suspect that stars and galaxies generate EM fields with periods of years, centuries, or millennia.*

The EM Wavelength Scale

We can illustrate the EM wavelengths with a *logarithmic nomograph*. We need the logarithmic scale because the wavelength range covers many *orders of magnitude* (powers of 10).

The left-hand portion of Fig. 11-6 is a logarithmic nomograph that shows wavelengths from 10^8 m down to 10^{-12} m. Each division, in the direction of shorter wavelength, represents a 100-fold decrease (two orders of magnitude). Utility AC shows up near the top of this scale; 60-Hz AC produces long EM waves. The gamma (γ) rays appear near the bottom of the nomograph; they have microscopic EM wavelengths.

Visible light, as we can see by looking at the left-hand nomograph, takes up only a tiny sliver of the whole range of EM wavelengths, known as the *EM spectrum*. In the right-hand scale, which expands the *visible spectrum*, we denote wavelengths in nanometers (nm).

PROBLEM 11-3

What's the wavelength of the 60-Hz EM field produced in free space by the AC in a common utility line?

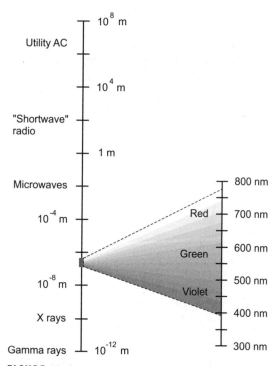

FIGURE 11-6 · At left, a logarithmic nomograph of the EM spectrum. At right, a magnified view of the visible-light range within the EM spectrum.

SOLUTION

We can use the formula for free-space EM wavelength (in meters) as a function of frequency (in hertz), obtaining

$$\lambda = 3.00 \times 10^8/f$$
$$= 3.00 \times 10^8/60$$
$$= 5.0 \times 10^6 \text{ m}$$

This wavelength represents approximately 5,000,000 m or 5000 km, half the distance from the earth's equator over the surface to the north geographic pole.

PROBLEM 11-4

Suppose that we drive variable-frequency AC through a wire conductor. As a result, we get an EM field around the wire, the wavelength of which varies as the AC frequency varies. Illustrate, by means of a generalized graph, how the AC frequency and the free-space EM wavelength relate.

SOLUTION

The frequency varies inversely in proportion to the wavelength (that is, in proportion to the reciprocal of the wavelength). Figure 11-7 shows the general relation between free-space EM wavelength and AC frequency.

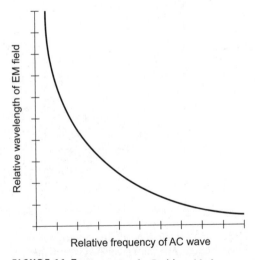

FIGURE 11-7 · Illustration for Problem 11-4.

PROBLEM 11-5

What's the frequency of the AC that produces an EM field with a wavelength of 75 m in free space? Express the frequency in hertz (Hz) and in megahertz (MHz).

SOLUTION

Let's use the formula for frequency (in hertz) as a function of wavelength (in meters), plugging in the numbers to get

$$f = 3.00 \times 10^8 / \lambda$$
$$= 3.00 \times 10^8 / 75$$
$$= 0.040 \times 10^8$$
$$= 4.0 \times 10^6 \text{ Hz}$$
$$= 4.0 \text{ MHz}$$

Electromagnetic Interference

When the EM fields from different electrical and electronic devices upset each other's operation, we get *electromagnetic interference* (EMI). In recent decades, EMI problems have worsened because consumer electronic devices have proliferated, and they've also grown more susceptible to EMI. Wireless devices are particularly vulnerable, both as "perpetrators" and as "victims."

Computers, Hi-Fi, and TV

Much of the EMI that plagues consumer devices occurs because of inferior equipment design. Faulty installation also contributes to the problem. A personal computer's *central processing unit* (CPU) produces wideband radio-frequency (RF) energy in the form of an EM field. The use of a CRT display compounds the trouble because CRTs generate EM fields as a result of the activity of their deflection coils. The digital pulses in the CPU and *peripherals* (such as keyboards, pointing devices, printers, and scanners) can also cause problems. The EM fields radiate from unshielded interconnecting cables and power cords because these cables and cords act as miniature transmitting antennas, as shown in Fig. 11-8A.

Computers, television (TV) receivers, and high-fidelity (hi-fi) sound equipment can malfunction because of RF fields from a nearby radio or television

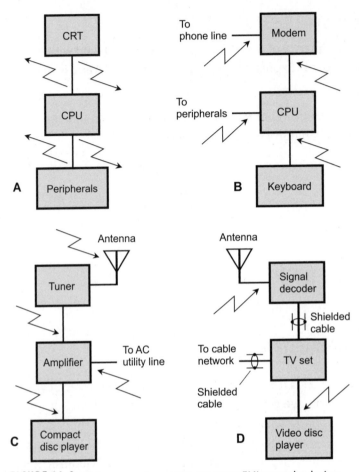

FIGURE 11-8 · At A, a computer system can cause EMI to nearby devices. At B, a computer system can suffer from EMI. At C, a hi-fi system can pick up EM fields. At D, shielded cables offer some protection against EMI in a home TV receiving system.

transmitter. In many cases, you can trace the problem to inadequate *RF shielding* in some part of the home-entertainment system or computer. In a computer system, cables and cords can act as receiving antennas (Fig. 11-8B), thereby letting RF energy into the CPU and causing the microprocessor to misbehave or "freeze up." In hi-fi sound equipment, RF can enter through unshielded speaker wires, the power cord, the antenna and feed line, and unshielded cables between an amplifier and externals, such as a compact disc (CD) player, as shown in Fig. 11-8C. In TV receiving installations, RF can enter through the power cord, the cable system, a satellite antenna system, and cords between the TV set and peripheral equipment, such as a video disc player (Fig. 11-8D).

Shielded cables, such as those shown in Fig. 11-8D between the TV set and the cable network, and between the TV set and the signal decoder, suffer less from EMI than unshielded cords and cables do.

In general, as you increase the number of interconnecting cables in a home entertainment system, the system becomes more susceptible to EMI. Also, as you lengthen the interconnecting cables, the potential for trouble increases. Good engineering practice dictates the use of as few connecting cords as possible, and keeping them as short as possible, in all home entertainment systems. If you have excess cord or cable and you don't want to cut it in order to shorten it, coil up the slack and tape it in place.

TIP *A good ground connection constitutes a critical design feature in any electronic home-entertainment system. Besides reducing your personal risk of electrocution, proper grounding can reduce the susceptibility of the system to EMI.*

A Special Note for Radio Hobbyists

If you enjoy *amateur radio* ("ham radio") or *shortwave listening* (SWLing), then you have experienced, or will experience, EMI-related radio reception problems. If you are a radio "ham" with a sophisticated station, you might find yourself taking the blame for interference to home entertainment equipment, whether it's technically your fault or not.

Amateur radio operators have been held responsible for imaginary "gremlins" that would astound competent engineers. But under certain circumstances, the RF fields from amateur radio transmitters can, in fact, present a danger. High-power transmitting antennas should be located so that humans can *never* come into direct contact with them. At microwave frequencies, the RF fields emitted by high-power transmitters connected to directional antennas can cause deep burns at close range. (The biological effect, if any, of long-term exposure to low-intensity RF fields remains unknown.)

When EMI originates from a broadcast station, an amateur radio station, a *citizens-band* (CB) radio station, a cell phone, a wireless utility *smart meter*, or any other wireless transmitting device, we call it *radio-frequency interference* (RFI). The fault for this type of interference rarely lies in the radio transmitter, which merely performs its job—the generation and emission of RF energy. Nevertheless, RFI from amateur or CB radio equipment can often be eliminated if the radio operator is willing to reduce the transmitter RF power output, change the operating frequency, or operate when neighbors don't use their home entertainment equipment.

PROBLEM 11-6

How can we make the arrangement shown in Fig. 11-8C less susceptible to RFI from nearby radio transmitters?

✔ SOLUTION

Figure 11-8C shows a hi-fi system in which none of the interconnecting cables have shielding. If we shield all of those cables, the system won't suffer as much from the RF fields that originate in nearby radio transmitters. The cable shields (usually comprising wire braid surrounding the internal conductors) should all be connected to the metal chassis of the amplifier. The amplifier chassis should, in turn, have a good connection to a substantial RF ground.

EM versus ELF

Many electrical and electronic devices produce EM fields with wavelengths much longer than the wavelengths of radio and wireless signals. Such EM fields arise from AC having an *extremely low frequency* (ELF). That's how the term *ELF fields* originated.

What ELF Fields Are (and Aren't)

The low-frequency limit of the *ELF spectrum* lies around 1 Hz. From there, the ELF range extends upward to approximately 3 kHz. This frequency range corresponds to free-space EM wavelengths exceeding 100 km. The most common ELF fields in today's environment have a frequency of 60 Hz. They emanate from utility wires in the United States and many other countries. (In some countries, it's 50 Hz.)

The military maintains land-based ELF installations that facilitate communications with submarines. These ELF waves propagate mainly beneath the earth's surface. Alternating currents at ELF travel underground and underwater more efficiently than AC waves at higher frequencies do. Some of this ELF energy "leaks" into the atmosphere near the earth's surface.

The term "ELF radiation" has caused confusion and unwarranted fear among lay people. The word "radiation" suggests grave, immediate danger that ELF fields simply do not pose, except in cases of extreme intensity never encountered in everyday life. The media attention to ELF fields has led some people to purchase devices that supposedly offer personal protection from the "threat," but in fact, do little or nothing.

? Still Struggling

An ELF field doesn't act like a barrage of X rays or gamma rays, which can cause sickness and death in large doses. Neither does ELF energy resemble UV radiation (which has been linked to skin cancer) or IR radiation (which can cause burns). An ELF field will not make anything radioactive. Some medical specialists have suggested that long-term exposure to high levels of ELF correlates with an abnormally high incidence of certain health problems. If you're worried about ELF fields, consult a qualified electrical engineer or health professional (or both).

ELF Fields and Computer Displays

As an ELF source, electromagnetic CRT displays have received particular attention. A conventional CRT creates images when electron beams strike and scan a phosphor coating on a glass screen. The electrons change direction as they sweep from left to right, and from top to bottom, on the screen. The sweeping is caused by deflection coils that steer the beam across the screen. The coils generate fluctuating magnetic fields that interact with the electrons, giving rise to EM fields at frequencies from a few hertz up to a few kilohertz.

Because of the positions of the coils and the shapes of the fields surrounding them, more ELF energy comes from the sides of a CRT display cabinet than from the front. If any true health hazard exists, it's greatest for someone sitting off to the side of a CRT display, and least for someone looking at the screen from the front.

The best "shielding" from ELF energy is physical distance, especially for people sitting near either side of a CRT display. The ELF field dies off rapidly with distance from the display cabinet. Computer workstations that use CRTs in an office environment should be at least 1.5 m (about 5 ft) apart. You should keep at least 0.5 m (about 20 in) away from the front of your own display. You can shut off the display when you're not using the computer. Most computers have a "power-saving" feature that shuts off the display automatically after a certain continuous period of nonuse.

TIP *Special electrostatic CRT displays, designed to minimize the generation of ELF fields, are available. They cost a lot of money, but they can offer peace of mind for people concerned about possible long-term health effects from exposure to ELF fields. The newer liquid-crystal displays (LCDs), which have largely replaced CRT displays in recent years, emit essentially no ELF energy.*

 PROBLEM 11-7

What's the wavelength of an ELF field that results from AC at a frequency of 10.0 Hz? Express the answer in meters (m).

✔ SOLUTION

To calculate this wavelength, we can use the formula stated earlier in this chapter for wavelength in terms of frequency, getting

$$\lambda = 3.00 \times 10^8 / f$$
$$= 3.00 \times 10^8 / 10.0$$
$$= 3.00 \times 10^7 \text{ m}$$

We'd have to travel three-quarters of the way around the world to go that far!

QUIZ

This is an "open book" quiz. You may refer to the text in this chapter. You'll find the correct answers listed in the back of the book.

1. An EM field with a free-space wavelength of 600 nm falls into the
   A. microwave spectrum.
   B. X-ray spectrum.
   C. visible spectrum.
   D. gamma-ray spectrum.

2. If we want to use an oscilloscope to look at a signal waveform, we normally apply that signal to the
   A. vertical deflection coils.
   B. anodes.
   C. electron gun.
   D. cathodes.

3. As an EM field propagates through space, the electric and magnetic lines of flux at any given point always
   A. run parallel to each other.
   B. run perpendicular to each other.
   C. rotate in opposite senses.
   D. rotate in the same sense.

4. Suppose that a galvanometer coil carries 40 mA of DC, causing the compass needle to deflect 32° to the east of geomagnetic north. What will the needle do if we reverse the direction in which the current flows, and decrease it to only 10 mA?
   A. The needle will deflect to the west of geomagnetic north by 32°.
   B. The needle will deflect to the west of geomagnetic north by some angle smaller than 32°.
   C. The needle will deflect to the east of geomagnetic north by some angle greater than 32°.
   D. The needle will deflect to the east of geomagnetic north by some angle smaller than 32°.

5. In free space, an EM wave at a frequency of 1.50 MHz has a wavelength of
   A. 600 m.
   B. 200 m.
   C. 150 m.
   D. 100 m.

6. Which of the following characteristics of a home-entertainment system could help to reduce or minimize its vulnerability to EMI effects?

   A. The use of long connecting cables
   B. The use of shielded connecting cables
   C. The absence of an electrical ground
   D. All of the above

7. An EM field can arise directly from

   A. a fixed permanent magnet.
   B. a fixed object carrying an electrical charge.
   C. constant DC passing through a straight wire.
   D. None of the above

8. The anodes in a CRT serve to

   A. accelerate and focus the electron beam.
   B. emit electrons and "shoot" them toward the cathodes.
   C. change the direction of the electron beam.
   D. prevent excessive output from the electron gun.

9. The Ångström is a unit of wavelength that represents

   A. 0.01 nm.
   B. 0.1 nm.
   C. 10 nm.
   D. 100 nm.

10. In free space, an EM wave measuring 12 m long has a frequency of

    A. 3.6 GHz.
    B. 2.5 GHz.
    C. 72 MHz.
    D. 25 MHz.

chapter 12

Practical Magnetism

Magnetism governs the operation of many devices and systems in our everyday lives. Let's look at some common examples.

CHAPTER OBJECTIVES

In this chapter, you will

- Analyze the operation of DC and AC electromagnets.
- See how certain materials intensify magnetic fields, while other materials have no effect and a few substances weaken them.
- Contrast the properties of permeability and retentivity.
- Learn how various everyday magnetic devices work.
- Discover how magnetic levitation can take place in the "real world," and how we can put it to constructive use.

Electromagnets

When we place a rod made of ferromagnetic material, called a *magnetic core*, inside a current-carrying coil of wire, as shown in Fig. 12-1, we get an *electromagnet*. The magnetic flux produced by the current temporarily magnetizes the core material. In addition, the magnetic lines of flux concentrate in the core. If the current grows large, the field strength in and near the core can rise to considerable levels.

DC Electromagnets

You can build a *DC electromagnet* by taking a large steel bolt and wrapping a couple of hundred turns of wire around it. You can find these items in a good hardware store. Make certain that the bolt is made of ferromagnetic material (but don't use a permanent magnet as your electromagnet core). Ideally, the bolt should measure at least 3/8 in (about 1 cm) in diameter, and at least 4 in (10 cm) long. You must use insulated or enameled wire (not bare wire), preferably made of solid, soft-drawn copper. So-called *bell wire* works well. Wind all of the wire turns in the same direction around the core.

A 6-V lantern battery can provide plenty of current to operate the electromagnet. When using a battery-powered electromagnet, don't leave the coil connected to the battery for more than a few seconds at a time.

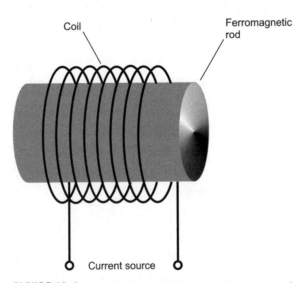

FIGURE 12-1 · A simple electromagnet comprises a wire coil surrounding a rod made of ferromagnetic material.

Direct-current electromagnets have defined north and south poles, just as permanent magnets do. However, an electromagnet can get much stronger than any permanent magnet can. Also, in an electromagnet, the magnetic field exists only while the coil carries current. When you power-down the device, the magnetic field collapses and disappears almost completely. A small amount of *residual magnetism* remains in the core of a DC electromagnet after powering-down, but the residual field is much weaker than the field generated when current flows in the coil.

WARNING! *Never use an automotive battery to power-up an electromagnet. The short circuit produced by the coil will draw a massive current, heating the battery acid and causing it to boil out explosively and burn you. Clothes offer no protection from this powerful acid, as I once learned from a painful experience. If the acid gets in your eyes, it can blind you permanently. Let automotive technicians work with (and worry about) "car and truck batteries." You stay away from them!*

AC Electromagnets

Do you imagine that you might make an immensely powerful electromagnet if, rather than using a lantern battery for the current source, you use utility AC straight out of a wall outlet? In theory, this scheme can actually work. But don't try it. Like the "car and truck battery" gamble, making an electromagnet with "wall current" is dangerous.

Some commercially manufactured AC electromagnets operate from household utility power, but they're designed with safety in mind. They incorporate devices such as step-down transformers, current limiters, and circuit breakers. These magnets "stick hard" to ferromagnetic objects. But they don't have constant north or south magnetic poles. The polarity of the magnetic field reverses every time the current reverses.

In an electromagnet operating from AC in the United States, 120 polarity reversals, or 60 complete north-south-north polarity cycles, take place every second. If we could measure and plot the magnetic-field polarity and intensity as a function of time, we'd get a graph something like Fig. 12-2.

WARNING! *Never plug an electromagnet into a wall outlet in an attempt to get a super-strong magnet. In the best case, it will instantly blow the fuse or circuit breaker at your utility panel. In the worst case, it will cause a fire in your home electrical wiring (perhaps inside the wall where you won't discover it until your whole house is burning down). Such an action also exposes you to the risk of electrocution.*

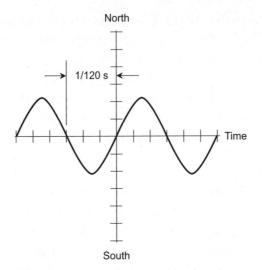

North

1/120 s

Time

South

FIGURE 12-2 • The polarity of an AC electromagnet reverses every 1/120 of a second when the AC frequency equals 60 Hz.

 PROBLEM 12-1

Suppose that the frequency of the current applied to a well-designed commercial AC electromagnet equals 50 Hz instead of the nominal 60 Hz commonly used in the United States. What will happen to the interaction between the alternating magnetic field and a nearby ferromagnetic substance such as iron?

SOLUTION

Assuming no change in the core material, the situation will remain essentially the same as it would be at 60 Hz. An AC electromagnet will function equally well at 50 Hz or 60 Hz.

Magnetic Materials

As we've learned, *ferromagnetic* materials, such as iron, nickel, or alloys containing either of these elements, compress magnetic flux lines considerably compared to their state in free space (a vacuum). We can permanently magnetize samples of such substances. Some materials, such as lithium, magnesium, molybdenum, and tantalum, compress magnetic flux lines a little, but we can't permanently magnetize them; physicists call them *paramagnetic*. *Nonferromagnetic* materials have essentially no effect on magnetic flux. A few materials, which physicists call *diamagnetic*, dilate the flux lines compared to their state in free space. Examples include wax, dry wood, bismuth, and silver. We can quantify the magnetic characteristics of any substance in terms of two characteristics: *permeability* and *retentivity*.

Permeability

Permeability, symbolized by the lowercase Greek mu (μ), is expressed and measured on a scale relative to free space. In theory, permeability values can range from 0 to "infinity" (arbitrarily large values). By convention, a perfect vacuum has $\mu = 1$. If we drive an electric current through a wire coil containing and surrounded by dry air, then the flux density in and around the coil is essentially the same as it would be in a vacuum. The permeability of dry air equals 1 for practical purposes, although in theory, it's a tiny bit larger. If the air contains a lot of water vapor (the humidity is high), the permeability increases a little more, but we can still consider it as equal to 1 in most situations.

If we place a core made of a ferromagnetic material, such as iron, nickel, or steel inside a coil, the flux density increases compared with the flux density in free space. Some substances make the flux density thousands of times greater than it would be in air or a vacuum with the same coil. If we use certain special metallic alloys as the core materials in electromagnets, the flux density, and therefore, the local strength of the field, can increase by a gigantic factor. Once in awhile you'll hear or read about a substance in which μ equals 10,000 (10^4) or 100,000 (10^5) or even 1,000,000 (10^6).

Diamagnetic materials actually spread out the magnetic lines of flux passing through them. In that sense we might call them "antimagnetic." A diamagnetic sample has a permeability value less than 1, but only a little less under ordinary circumstances. Engineers and technicians use diamagnetic objects to keep magnets physically separated while minimizing the interaction between them. Table 12-1 lists some common materials and their approximate permeability values.

TIP *Under extreme low-temperature conditions, good electrical conductors can lose all their resistance. When this happens, they attain a theoretical permeability of 0. Such a material, known as a* **superconductor,** *actually expels magnetic flux, so a magnetic field can't exist inside it at all! Under the right conditions, superconductors produce* **magnetic levitation,** *as we'll see later in this chapter.*

TABLE 12-1 Magnetic permeability factors for some common materials. All figures are approximate except the value for free space (a vacuum), which equals exactly 1 by convention.

Substance	Permeability
Air, dry	Slightly more than 1
Aluminum	Slightly more than 1
Bismuth	Slightly less than 1
Cobalt	60 to 70
Ferrite	100 to 3000
Free space (vacuum)	1 (exactly)
Iron	60 to 8000
Nickel	50 to 60
Permalloy	3000 to 30,000
Silver	Slightly less than 1
Steel	300 to 600
Specialized alloys	100,000 to 1,000,000

Retentivity and Core Saturation

When we subject a ferromagnetic sample to a magnetic field by enclosing it in a wire coil carrying a high current, we'll observe some residual magnetism when the current stops flowing in the coil. *Retentivity*, also called *remanence*, quantifies the extent to which a substance can "memorize" a magnetic field imposed on it, thereby becoming a permanent magnet.

If we construct an electromagnet with a certain core material and then drive increasing DC through its coil, the flux density inside the core goes up for awhile. As we increase the current beyond a certain point, however, the flux density levels off. Past that limit, increases in the coil current yield no further increase in the core flux density. Once we've forced a DC electromagnet's core to hold as much flux density as it possibly can while a current flows in the coil, we have a condition called *core saturation*. In order to determine the retentivity for the core material, we must compare the flux density under conditions of core saturation to the residual-field flux density that remains in the core after we cut off the current in the coil.

Suppose that for a particular substance, the maximum possible flux density, called the *core-saturation flux density*, equals x tesla or gauss. Suppose the flux density in the core diminishes to y tesla or gauss after we remove the current from the coil. We'll find that y equals a fraction of x. We define the retentivity, B_r, of the material as

$$B_r = y/x$$

We can also express the retentivity as a percentage $B_{r\%}$, where

$$B_{r\%} = (100\ y/x)\%$$

Imagine that we can magnetize a metal rod to a flux density of up to 1000 G, but no more than that, when we surround it with a coil carrying DC. For any substance, such a maximum always exists; further increasing the current in the wire can't make the flux density within the core grow any greater. Now suppose that we remove the current from the coil, and the flux density in the rod drops to 20 G. We can calculate the retentivity as

$$B_r = 20/1000$$

$$= 0.020$$

As a percentage, we have

$$B_{r\%} = (100 \times 20/1000)\%$$

$$= 2.0\%$$

TIP   *Certain ferromagnetic substances have high retentivity. They make excellent choices for the manufacture of permanent magnets. Other ferromagnetic materials have low retentivity. They work as electromagnet cores, but they don't make good permanent magnets.*

?

Still Struggling

Sometimes you'll want a substance with good ferromagnetic properties but low retentivity. For example, suppose that you want to construct an electromagnet that will operate from DC so that it maintains constant polarity, but that loses its magnetism after you remove the current. If a ferromagnetic substance has low retentivity, it works well as the core for an AC electromagnet because the polarity of the magnetic flux can easily reverse. If the core retentivity is high, the material exhibits "magnetic sluggishness"; its flux can't follow rapid current reversals in the coil. That's why a substance with high retentivity won't work well as the core material in an AC electromagnet.

PROBLEM 12-2

Suppose that you increase the frequency of the current applied to an AC electromagnet to some value far above 60 Hz. Imagine, for example, that you have a variable-frequency AC electrical source of constant voltage, capable of delivering constant current. You increase the frequency to 600 Hz, then 6 kHz, then 60 kHz, 600 kHz, 6 MHz, 60 MHz, 600 MHz, and all the way up to 6 GHz! What will happen to the magnetic field as the frequency increases?

✔ SOLUTION

The electromagnet will work normally as the frequency increases, until the coil-and-core combination begins to impede the flow of AC. Then the magnetic field strength will get smaller and smaller, gradually tapering off to near zero as the frequency continues to rise. The electromagnet's opposition to high-frequency AC occurs in part because of an increase in the tendency of all coils to impede the flow of AC at high frequencies (a phenomenon called *reactance*). In addition, the core retentivity has an increasingly detrimental effect on the peak magnetic field strength as the AC frequency increases. At the highest frequencies, the coil in the electromagnet will act like a large-value resistor, and the core material won't have enough time to get magnetized at one polarity before the current reverses, thereby reversing the polarity of the magnetizing flux.

PROBLEM 12-3

Suppose that you surround a metal rod with a coil. You can make the magnetic flux density as great as 0.4567 T inside the rod, but further increases in current cause no more increase in the flux density inside the rod. Then you remove the current from the coil, and the flux density inside the rod drops to 514 G. What's the retentivity of this core material, expressed as a percentage?

✔ SOLUTION

Before doing any calculation, you must convert both flux density figures to the same units. Remember that $1 \text{ T} = 10^4 \text{ G}$. Therefore, the flux density equals $0.4567 \times 10^4 = 4567$ G with the current. You know that it's 514 G after you remove the current. "Plugging in" these numbers tells you that

$$B_{r\%} = (100 \times 514/4567)\%$$
$$= 11.3\%$$

PROBLEM 12-4

What's the retentivity, in general, of a diamagnetic material, such as wax or dry wood? How about dry air?

SOLUTION

Any material that has permeability of 1 or less does not concentrate magnetic lines of flux. If the core inside a coil comprises a nonferromagnetic material, it doesn't retain any magnetic flux after the current is removed from the coil. Therefore, the retentivity equals 0 (or 0%).

Permanent Magnets

You can use any ferromagnetic material, or any substance whose atoms can maintain their alignment, to make a permanent magnet. You played with permanent magnets as a child, curious as to which objects would "stick" to them and which things wouldn't; today you use permanent magnets to attach grocery shopping lists to your refrigerator! (Where did all the fun go?) As you can guess, some alloys produce stronger permanent magnets than others.

Permanent magnets work best when made from materials with high retentivity. You can permanently magnetize a sample of such material by using it as the core of a DC electromagnet for an extended period of time. Weak permanent magnets can be made using other permanent magnets. For example, if you want to magnetize a screwdriver with a steel shaft so that it will hold onto screws when inserting them in hard-to-reach places, you can slide one end of a strong bar magnet about 100 times along the screwdriver's shaft from the handle to the tip.

TIP *Once you've magnetized a metal tool, it will "stay magnetic" to some extent indefinitely. Make sure that you want your tool to become a permanent magnet, however weak, before you magnetize it!*

Flux Density inside a Long Coil

Consider a long coil of wire (commonly known as a *solenoid*) that has *n* turns, and whose length in meters equals *s*. Suppose that this coil carries a direct current of *I* amperes, and has a core whose permeability equals μ. We can calculate the flux density B_t (in teslas) inside the core, assuming that it's not driven into saturation, using the formula

$$B_t = 4\pi \times 10^{-7} \ (\mu nI/s)$$

A good approximation is

$$B_t = 1.2566 \times 10^{-6} \ (\mu nI/s)$$

PROBLEM 12-5

Suppose that a DC electromagnet carries a certain current. It measures 20 cm long, and has 100 turns of wire. The flux density in the core, which is not in a state of saturation, equals 20 G. The permeability of the core material equals 100. How much current, in milliamperes, flows in the wire, accurate to the nearest milliampere?

SOLUTION

First, we must convert the units to the correct forms as necessary. The length *s* is 20 cm, equivalent to 0.20 m. The flux density equals 20 G, so we know that B_t = 0.0020 T. Now let's rearrange the above formula for flux density so that we get a formula for *I* instead. We start with

$$B_t = 1.2566 \times 10^{-6} \ (\mu nI/s)$$

Using some high-school algebra, we can "morph" it into

$$I = 7.9580 \times 10^5 \ (sB_t/\mu n)$$

When we input the known values, we get

$$I = 7.9580 \times 10^5 \ (0.20 \times 0.0020)/(100 \times 100)$$
$$= 7.9580 \times 10^5 \times 4.0 \times 10^{-8}$$
$$= 0.031832 \text{ A}$$
$$= 31.832 \text{ mA}$$

We can round this result off to 32 mA.

Magnetic Devices

Let's look at some machines that take advantage of permanent magnets or electromagnets to perform specific tasks. Because all of these devices convert electrical energy into mechanical work or vice-versa, we can classify them as *electromechanical transducers*.

The Chime

Figure 12-3 is a simplified diagram of a bell ringer, also called a *chime*. Its solenoid constitutes an electromagnet. The core has a hollow region in the center, along its axis, through which a steel rod passes. The coil contains many turns of wire, so the electromagnet can produce a powerful field if a substantial current passes through. When no current flows in the coil, gravity holds the rod (called the *hammer*) down against the plastic base plate. When we drive a surge of current through the coil, the magnetic field pulls the hammer upward. The magnetic field "tries" to align the ends of the hammer with the ends of the coil core (and if the current flowed continuously and indefinitely, that would happen).

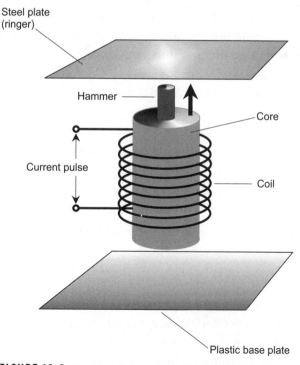

FIGURE 12-3 · Functional diagram of a solenoid that works as a bell ringer.

But the current pulse lasts for only a fraction of a second. By the time the hammer has risen to the point where it lines up with the core, the current pulse has ended, and the magnetic field has disappeared. The momentum of the hammer, therefore, carries it upward until it strikes the steel plate (called the *ringer*) and then bounces back down to rest on the plastic base again. The impact of the hammer against the ringer produces the chime sound.

The Relay

In some electrical and electronic systems, you can't conveniently place a switch at the optimum location. For example, you might want to switch a communications line from one branch to another from a long distance away. In wireless transmitters, some of the wiring carries high-frequency AC that must remain confined to certain parts of the circuit. A *relay* makes use of a solenoid to allow remote-control switching.

Figure 12-4A is a functional drawing of a relay, and Fig. 12-4B shows the schematic symbol for the same relay. Engineers call this particular device a *single-pole double-throw (SPDT) relay* because it can switch a single terminal between two different circuits. The movable lever, called the *armature*, is held to one side by a spring or by its own tension when no current flows through the electromagnet. Under these conditions, terminal X connects to Y, but not to Z. When we apply a sufficient voltage to the coil to force current through it, the armature is pulled down by the magnetic force exerted on a ferromagnetic disk by the solenoid. This motion disconnects terminal X from Y, and connects X to Z.

Some relays are designed for use with DC, and others are intended for switching AC. Some can work with either AC or DC. A *normally closed relay* completes the circuit when no current flows in its solenoid, and breaks the circuit when current flows. A *normally open relay* completes the circuit when current flows in its solenoid, and breaks the circuit when current does not flow. (In this context, the "normal" condition refers to no current in the solenoid.) The relay shown in Fig. 12-4 can function as a normally open relay or as a normally closed relay, depending on which contacts we connect to the external circuits. This relay can also switch a single line between two different circuits.

TIP *Nowadays, you'll find relays only in systems carrying high currents or voltages, or where conditions demand exceptional ruggedness. In most applications, semiconductor switches work better than relays do. Semiconductor devices (such as diodes and transistors) have no moving parts, so they can perform many more switching operations per second than relays can.*

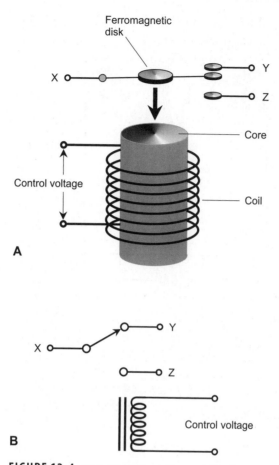

FIGURE 12-4 • At A, a functional diagram of a relay. At B, the schematic symbol.

The DC Motor

Magnetic fields can be harnessed to do mechanical work. A *DC motor* converts DC energy into rotational mechanical energy. Some motors have dimensions so small that we need a microscope to see them. Others are so large that we need heavy equipment to move them or transport them. Some tiny motors can function in medical devices that circulate in the bloodstream or help body organs to carry on their work. Massive motors can pull trains at hundreds of kilometers per hour. (Most motors have size and weight somewhere in between these extremes, of course!)

In a DC motor, the source of electricity connects to a set of coils, producing magnetic fields around the coils. The attraction between opposite poles, and the repulsion between like poles, is switched in such a way that a constant *torque*,

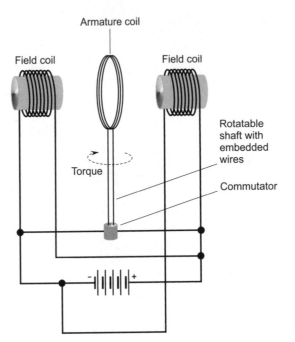

FIGURE 12-5 · Functional diagram of a DC motor.

or rotational force, results. As we increase the coil current, the mechanical torque increases, up to a certain maximum that represents the motor's mechanical-power-output limit.

Figure 12-5 is a simplified functional diagram of a DC motor. The *armature coil* rotates with the motor shaft. A stationary pair of *field coils* is affixed to the motor housing. (In some motors, a pair of permanent magnets replaces the field coil.) The *commutator* reverses the current direction in the armature coil with every half-rotation of the motor shaft. This action keeps the shaft's rotational force going in the same sense, either clockwise or counterclockwise. The shaft must have enough mass so that its rotational momentum keeps it from stopping at the instants in time when the commutator reverses the current.

The Stepper Motor

A *stepper motor* turns in small increments, rather than continuously. The *step angle*, or extent of each turn, varies depending on the particular motor. It can range from less than 1° of arc to a quarter of a circle (90°). A stepper motor turns through its prescribed step angle and then stops, even if the current is maintained. In fact, when a stepper motor comes to rest with current going through its coils, the shaft resists applied torque.

Conventional motors run at hundreds, or even thousands, of revolutions per minute (RPM). A stepper motor usually runs at less than 180 RPM, and often much less. In a conventional motor, the torque increases as the motor runs faster. But in a stepper motor, the torque decreases as the motor speed increases. A stepper motor has the most turning power when it is running at slow speed. When it's at rest but carries current, it tends to stay in place and resist applied rotational forces. In effect, it has electromagnetic brakes!

The most common stepper motors can be found in either of two types. A *two-phase stepper motor* has two coils, called *phases*, controlled by four wires. A *four-phase stepper motor* has four phases and eight wires. The motors are advanced incrementally (stepped) by applying currents to the phases in a controlled sequence. Figure 12-6 shows schematic diagrams of two-phase (at A)

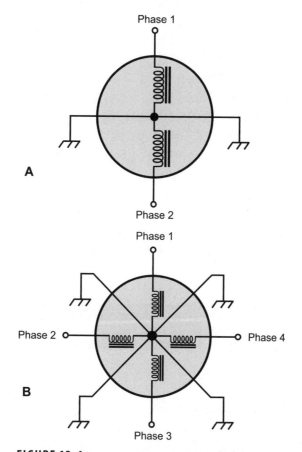

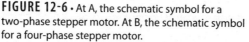

FIGURE 12-6 • At A, the schematic symbol for a two-phase stepper motor. At B, the schematic symbol for a four-phase stepper motor.

and four-phase (at B) stepper motors. Table 12-2 shows control-current sequences for two-phase (at A) and four-phase (at B) stepper motors.

When we apply a pulsed current to a stepper motor, with the current rotating through the phases, as shown in the tables, the motor rotates in increments, one step for each pulse. In this way, the motor maintains a constant and precise rotational speed. Because of the braking effect, this speed remains the same over a wide range of mechanical loads (turning resistances). This mode contrasts with a conventional DC motor, which tends to slow down when the mechanical load increases, or to speed up when the mechanical load decreases or disappears.

The Selsyn and Synchro

A *selsyn* uses stepper motors to indicate the direction in which an object points or aims. A transmitting motor connects to the movable object, and also to a receiving motor at a convenient location. The transmitting motor senses the orientation of the object, and the receiving motor, equipped with a compass-like indicating needle, displays this orientation. If you have a rotatable communications antenna system, a selsyn most likely serves as the direction indicator.

In a selsyn, the indicator rotates the same number of degrees as the moving device. A selsyn designed to indicate the azimuth, or horizontal compass bearing, can rotate through a range of 360°. A selsyn designed to indicate elevation, or the angle with respect to the horizon, can rotate through a range of 90° for use on the earth's surface, or 180° for use in space.

A *synchro* is a special type of stepper motor, used for remote control of mechanical devices. A synchro consists of a generator and a receiver motor. As a control operator turns the generator shaft, the receiver motor's shaft follows

TABLE 12-2A	Operation of a two-phase stepper motor. Read down for clockwise rotation. Read up for counterclockwise rotation.	
Step	**Phase 1**	**Phase 2**
1	off	off
2	on	off
3	on	on
4	off	on

TABLE 12-2B Operation of a four-phase stepper motor. Read down for clockwise rotation. Read up for counterclockwise rotation.

Step	Phase 1	Phase 2	Phase 3	Phase 4
1	on	off	on	off
2	off	on	on	off
3	off	on	off	on
4	on	off	off	on

along exactly. A synchro acts like a powerful selsyn in reverse. Synchros have many uses, especially in robotics. They're ideal for controlling fine motion, and also for robotic *teleoperation* (remote control).

TIP *Several stepper motors or synchros, all under the guidance of a program-mable chip called a* **microcontroller,** *can provide precision control of robotic devices. Sets of stepper motors or synchros work particularly well for* **point-to-point motion,** *in which the coordinates of a robotic device's* **end effector (hand, gripper, or specific tool)** *in space are precisely "mapped" by a microcontroller as a function of time.*

The Fluxgate Magnetometer

A *fluxgate magnetometer* comprises a computerized mobile-robot guidance system that uses magnetic fields to tell the robot its location and orientation. Navigation within a defined area can be carried out by having the *robot controller* (a microcomputer contained in the robot) constantly analyze the orientation and intensity of the lines of flux generated by fixed electromagnets. Figure 12-7 shows a simplified hypothetical scenario, with two fixed electromagnets and a robot in the magnetic field produced by them. In this case, opposite magnetic poles (north and south) face each other, giving the field the characteristic opposing-pole bar-magnet flux contour.

For each point in the robot work environment, the magnetic flux has a unique orientation and intensity. There exists a one-to-one correspondence between magnetic flux conditions and each point within the robot's work environment. The robot controller contains data and software so that it "knows" this relation precisely for all points in the vicinity. The relation and data, together

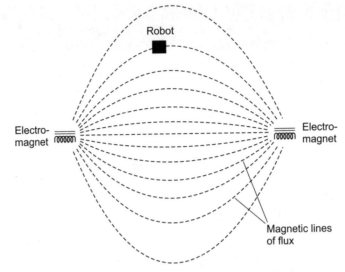

FIGURE 12-7 · A fluxgate magnetometer allows a mobile robot to determine its location, based on the strength and orientation of the flux lines produced by fixed electromagnets.

known as a *computer map*, allows the robot controller to determine the robot's location and orientation at all times. If the data is updated at frequent intervals, the velocity (speed and direction) at which the robot moves can also be determined and recorded.

PROBLEM 12-6

Can we employ the earth's magnetic field, rather than an artificially generated magnetic field, in a fluxgate magnetometer system for a mobile robot?

SOLUTION

Yes. In this case, the robot's work environment becomes, in effect, the entire surface of the earth, except for locations near the geomagnetic poles, or places where the earth's magnetic field cannot be sensed. But the robot will wander astray if an artificially generated magnetic field exists in addition to the earth's magnetic field, unless we program the robot to correct for interaction between the two magnetic fields. Similar "confusion" can occur for robots operating in concrete-and-steel buildings or in magnetically shielded enclosures because such structures can block or distort the earth's magnetic lines of flux.

The Dynamic Transducer

A *dynamic transducer* comprises a coil-and-magnet combination that translates mechanical vibration into varying electrical current or vice-versa. The most common examples are the *dynamic microphone* and the *dynamic speaker*.

Figure 12-8 is a functional diagram of a dynamic transducer. A diaphragm is attached to a coil that can move back and forth rapidly along its axis. A permanent magnet rests inside the coil. Sound waves cause the diaphragm and coil to move together, producing fluctuations in the magnetic field within the coil. As a result, audio AC flows in the coil, having the same waveform as the acoustic disturbance that strikes the diaphragm.

If we apply an audio signal to the coil, the AC in the wire generates a magnetic field that produces forces on the coil. These forces cause the coil to move, pushing the diaphragm back and forth to create acoustic waves in the surrounding air.

PROBLEM 12-7

What would happen if DC, rather than audio-frequency AC, were applied to the coil of a dynamic speaker?

✔ SOLUTION

A magnetic field would arise within the coil of wire, pulling it either in or out (depending on the direction of the current). The diaphragm would, therefore, be "sucked in" or "pushed out," but it would not vibrate. No sound would come from the speaker, except for a click when we first applied the DC, and another click when we removed the DC. It's a bad idea to apply very

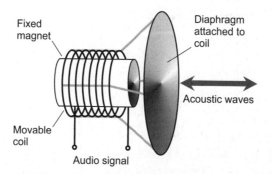

FIGURE 12-8 · Functional diagram of a dynamic transducer, such as a microphone or speaker.

much DC to a dynamic transducer, even in combination with an AC signal because the constant magnetic field can interfere with normal operation.

Magnetic Media

The term *magnetic media* refers to hardware devices and systems that allow the storage and recovery of data in the form of magnetic fields. In computers, the most common magnetic medium is the *hard disk*, also called the *hard drive*. The *magnetic diskette*, once popular for external data storage, has been replaced by optical media, such as *compact discs (CD)* and electronic memory modules known as *flash drives* or *thumb drives*. Some computer enthusiasts use magnetic *tape drives* to store large amounts of data for long periods of time. Some audio and video recording studios also employ magnetic tape. For consumer use, the *digital versatile disc (DVD)*, which resembles the CD but has more storage capacity, has supplanted magnetic tape.

How It Works

In magnetic media, millions of tiny ferromagnetic particles adhere to the surface of a tape or disk. Each particle can be magnetized and demagnetized repeatedly. These particles can maintain a given state of magnetization for a period of years (sometimes decades). Early in the twentieth century, audio engineers noticed this property and saw its potential for recording sound. As a result, the *audio tape recorder* came into existence. Around the middle of the twentieth century, engineers adapted audio tapes for use in the first computers. Since then, new forms of magnetic media have evolved, but the essential technology hasn't changed at all.

Magnetic data can be erased and overwritten thousands of times, but when left alone, a magnetic tape or disk constitutes a form of *nonvolatile data storage*. That means they don't require a constant source of power to maintain the information content. You can switch a computer or tape recorder off, and the disk or tape will retain the information that you've stored on it.

Magnetic disks enjoy one special advantage over their magnetic tape counterparts: You can write data onto, or read data from, a disk much faster than you can do either task with a tape. No two data elements are ever separated by more than the diameter of a disk (a few centimeters), while two different elements can be separated by the entire length of a tape (up to hundreds of meters).

Magnetic media exhibit troublesome heat sensitivity. If the temperature rises above roughly 100°F (38°C), the ferromagnetic atoms move around so fast that their electron orbits lose their alignment more rapidly than they do at room temperature. A gradual demagnetizing effect occurs regardless of the temperature, given enough time, but excessive heat makes it happen fast enough that it can cause problems. You should always store magnetic disks and tapes in a cool place. You should never leave them in direct sunlight, or inside an enclosed vehicle during sunny weather.

Magnetic media are sensitive to strong magnetic fields. You should keep disks and tapes away from permanent magnets, electromagnets, and from anything that generates magnetic fields. Loudspeakers, headphones, microphones, and the back ends of cathode-ray-tube (CRT) monitors emanate magnetic flux. Don't carry computer diskettes in a handbag or briefcase along with your "refrigerator magnets"!

TIP *Magnetic disks and tapes don't last forever, even if you provide ideal storage conditions. It's a good idea to renew everything (copy all your archives to new media) annually for computer disks, and every three to five years for magnetic tapes.*

Magnetic Tape Recording

Magnetic tape is commonly used for storing sound, video, and digital information. It's available in various thicknesses and widths for different applications. The ferromagnetic particles are attached to a long, thin, flexible, stretch-resistant strip or *substrate* made of strong plastic. In some cases, nonferromagnetic metal is used instead of plastic. Tapes from the "really olden days" (circa 1940s) had substrates made of paper! A magnetic field, produced by a *recording head*, causes polarization of the particles. As the field fluctuates in intensity, the polarization of the particles varies in strength and alternates back and forth. When the tape is played back, the magnetic fields surrounding the individual particles produce current changes in the *playback head*, also called the *pickup head*.

For sound and computer-data recording, magnetic tape is available in cassette form or reel-to-reel form. The tape thickness can vary. Thicker tapes have better resistance to stretching, although the recording time, for a given length of tape, is proportionately shorter than with thin tape.

Magnetic tape provides a convenient and compact medium for long-term storage of information. But certain precautions must be observed. The tape must be kept clean and free from grease. Magnetic tapes should be kept at a reasonable temperature and humidity, and should not be subjected to magnetic fields.

For illustrative purposes, let's look the operation of an old-fashioned analog tape recorder. All such devices act as transducers between acoustic, digital, or video signals and variable magnetic fields. Figure 12-9 is a simplified rendition of the recording and playback apparatus in a typical audio tape recorder, of the sort you can still buy today.

In the *record mode*, the tape moves past the *erase head* before anything is recorded. If the tape is not blank (that is, if magnetic impulses already exist on it), the erase head removes the impulses before anything else is recorded. This precaution prevents *doubling* (simultaneous presence of two programs on the tape). The *recording head* is an electromagnet that generates a fluctuating magnetic field, whose instantaneous flux density varies in proportion to the instantaneous level of the audio input signal. This fluctuation magnetizes the tape in a pattern that duplicates the waveform of the signal.

In the *playback mode*, the erase head and recording head are not activated, so they and their associated circuits become irrelevant. The *playback head* acts as a magnetic-field detector. As the tape moves past, the playback head is exposed to a fluctuating magnetic field whose waveform mimics the waveform that the recording head produced originally. This magnetic field induces AC in the playback head. The AC is amplified and delivered to a speaker, headset, or other output device.

The Magnetic Hard Drive

A magnetic hard drive is a common form of *mass storage* for computer data. The drive consists of several disks, called *platters*, arranged in a stack and made of rigid, durable material coated with a finely powdered ferromagnetic substance. The platters are spaced a fraction of a centimeter apart. Each has two

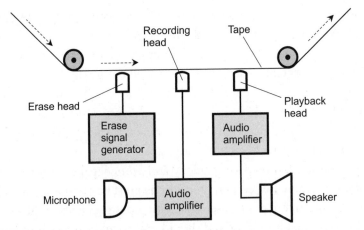

FIGURE 12-9 · Functional diagram of an old-fashioned magnetic tape recorder.

sides (top and bottom) and two *read/write heads* (one for the top and one for the bottom). A sealed cabinet holds the assembly.

When the computer is switched off, the hard drive mechanism locks the heads in a position away from the platters, preventing damage to the heads and platters in the event of mechanical shock. When the computer is powered-up, the platters spin at several thousand RPM. The heads hover a few millionths of a centimeter above and below the platter surfaces. When you type a command or "click" (activate) a display icon or button telling the computer to read or write data, the hard drive mechanism goes through a series of rapid, complicated, precise movements. The head positions itself over the spot on the platter where the data is located or is to be written; then the head detects the magnetic fields and translates them into tiny electric currents. This whole process takes only a small fraction of a second.

The data on a hard drive appears physically in concentric, circular *tracks*. Hundreds or thousands of tracks exist per radial centimeter of the platter surface. Each circular track comprises multiple arcs called *sectors*. A *cylinder* is the combination of equal-radius tracks on all the platters in the drive. Tracks and sectors are set up on the hard drive during the initial *formatting* process. The hard drive also contains data units called *clusters* consisting of one to several sectors, depending on the arrangement of data on the platters.

When you buy a new computer, it will likely have a hard drive built in. The drive comes installed and formatted. Most new computers are sold with several commonly used programs installed on the hard drive. Some computer users prefer to buy new computers with only the *operating system* (OS) installed, which all programs require to run properly. This "bare-bones" approach leaves plenty of hard-drive space available, and gives the user control over which programs to install (or not).

PROBLEM 12-8

Can the so-called "X-ray" scanners, commonly used at airports, erase the data on a magnetic disk or tape? What about the metal detectors?

SOLUTION

All by themselves, X rays can't affect magnetic disks or tapes, or do anything to the data written onto them. (However, X rays may erase or damage data on certain types of computer memory chips.) Walk-through and portable metal detectors generate magnetic fields, but some disagreement exists as to whether these fields attain flux-density levels strong enough to affect the data on a magnetic disk or tape. If you have any concerns, you can hand your tapes or diskettes to the attendants and let them return the magnetic media to you after you've walked through the scanner or metal detector. You can also send magnetic media by old-fashioned post or courier to your destination before you start your trip! As another alternative, you can store your data on optical media, such as CDs, which neither X rays nor magnetic fields can affect. In any case, you should keep redundant backups of important data at multiple locations to protect your information against loss, fire, theft, or any other misfortune.

Magnetic Levitation

The next time you have access to a couple of powerful permanent magnets, hold them so their matching poles are only a centimeter or two apart (north-to-north or south-to-south). As you push the magnets closer to each other, feel the way they push back against you. Do you remember the first time you observed this phenomenon? Did you notice that the repulsive force could exceed the weight of either magnet? Did this make you wonder if magnets could produce anti-gravity effects?

An Experiment

Imagine that you glue a couple of dozen small, pellet-shaped "refrigerator magnets" in a matrix all over the inside surface of a plastic mixing bowl, with the north poles facing upward. You get, in effect, a large permanent magnet with a concave, north-pole surface. Suppose that you anchor this bowl to a tabletop, and then take a single pellet-shaped magnet and hold it with its north pole facing downward, precisely over the center of the mixing bowl, as shown in Fig. 12-10A. As soon as you let go of the single magnet, it will flip over and "stick" to one of the magnets that you've glued inside the bowl.

Now suppose that you take another couple of dozen magnets and glue them to another mixing bowl of the same shape, but only about 3/4 the size of the first bowl. You glue the magnets all over the outside of the smaller bowl, with the north poles facing away from the bowl, thereby obtaining a large magnet with a convex north-pole surface. Suppose you try to set this bowl down inside the first one, as shown in Fig. 12-10B. Do you think that the top "bowl magnet" will hover above the bottom one? Well, it won't! It will find a way to "get off center" and come to rest against the lower bowl.

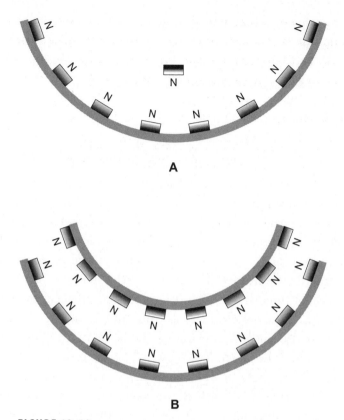

FIGURE 12-10 · When you try to levitate a magnet above a set of other magnets, as shown at A, the top magnet flips over and sticks to one of the others. Instability also occurs with two bowl-shaped magnetic structures, one above the other, as shown at B.

Earnshaw's Theorem

When pioneer magnetic-levitation seekers performed experiments, such as those described above and shown in Fig. 12-10, they might have thought that levitation, like perpetual motion, is impossible. In fact, magnetic levitation can't be achieved with a set of *static* (stationary and nonrotating) permanent magnets. Some instability always occurs in such a system, and the slightest disturbance magnifies the imbalance, ultimately causing the whole ensemble to "crash." A scientist named *Samuel Earnshaw* proved this fact with mathematical certainty in the 1800s. Today, we call his result *Earnshaw's Theorem*.

Despite the pessimistic conclusion of Earnshaw's Theorem, we can obtain magnetic levitation in the "real world" because the theorem rests on a narrow set of assumptions, and clever engineers can construct systems that get around these constraints. Earnshaw's Theorem applies only to sets of *permanent* magnets in arrangements where *no relative motion* occurs. In recent years, scientists have come up with dynamic (moving) systems of magnets that can produce levitation. Some such systems have been put to practical use.

Feedback Systems

Consider the two-bowl scenario of Fig. 12-10B. If you try this experiment, you'll get frustrated. But suppose you build a *feedback system* that keeps the upper bowl in alignment with the lower one? Feedback systems exist in all sorts of regulated devices, from the governors on motors to the oscillators in radio transmitters and receivers. Suppose, in addition, that you set the upper bowl spinning, creating a *gyroscope effect* that helps to keep it oriented properly in the horizontal plane.

Here's a crude example of how an electromechanical feedback system can produce magnetic levitation with the two-bowl scheme. It has an electronic *position sensor* that produces an *error signal*, and a mechanical *position corrector* that operates, based on the error signals from the position sensor, to keep the upper bowl from drifting off center. As long as the rotating upper bowl remains exactly centered over the lower one (Fig. 12-11A), the position sensor produces no output signal. The upper bowl has a tendency to move sideways because of instability in the system (Fig. 12-11B). As soon as the upper bowl gets a little off center, the position sensor produces error data that describes the extent and direction of the displacement. The error data has two components: a *distance error signal* that gets stronger as the off-center displacement of the upper bowl increases, and a *direction error signal* that indicates the compass bearing (or azimuth) in which the upper bowl has drifted. These signals are sent to a microcomputer that operates a mechanical device that produces the necessary amount of force, in exactly the right direction, to get the upper bowl back into alignment with the lower one (Fig. 12-11C).

? Still Struggling

Here's an analogy that can help you imagine how a feedback system of the above-described sort can work. Think about how you steer a car down the proper lane while driving on a highway. The car is inherently unstable. If you take your hands off the steering wheel for a moment, your car will veer out of its lane. But the car stays in its lane because you constantly make minor course corrections. You and the car together constitute a feedback system. If the car starts to drift toward the right, your eye and brain sense the problem, and error data goes to your hands. Your hands produce a slight counterclockwise force on the steering wheel, which operates a system that turns the wheels a little bit, steering the car back toward the left. If the car starts to drift toward the left, you sense it, your hands produce a slight clockwise force on the steering wheel, and the car veers back toward the right. A machine-vision system, some electronic circuits, and a microcomputer connected to the steering apparatus could substitute for you, and keep the car centered in its lane.

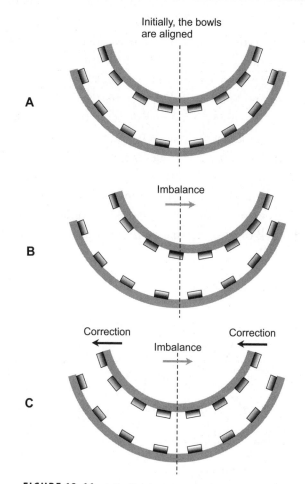

FIGURE 12-11 · A feedback system can keep two magnet-arrayed bowls centered, producing levitation, if the upper bowl rotates to produce a gyroscope effect. At A, the initial aligned situation. At B, an imbalance throws the upper bowl off center. At C, a correction force brings the upper bowl back into alignment.

Diamagnetic Materials

As we've learned, diamagnetic materials "spread out" magnetic lines of flux compared with the flux density in free space. These substances, under normal conditions, do not reduce the flux density nearly as much as ferromagnetic materials concentrate it, but diamagnetic materials can, under the right conditions, behave like "antimagnets."

A diamagnetic substance, such as dry wood or distilled water, repels either pole of a permanent magnet, just as a ferromagnetic substance attracts either pole. We don't notice this effect with ordinary magnets because they're not strong enough. When you, as a child, played with magnets, you didn't observe a force of repulsion against a hardwood floor, or against a piece of paper, or against a window pane. You didn't toss a permanent magnet at a pond and then watch it hover above the surface. It sank, of course, and water wasn't repelled from it, either. You noticed that some things attracted magnets and other things didn't, and you gave the matter no further thought. (Or did you?)

Repulsion between diamagnetic objects and magnets becomes significant if the magnets can produce intense enough magnetic fields. Levitation can occur if we place a lightweight diamagnetic object inside a bowl-shaped container arrayed with specially designed, high-current electromagnets. A small drop of distilled water, for example, can be suspended in midair. Experimenters have actually accomplished this feat in lab experiments, using an arrangement similar to the one shown in Fig. 12-12.

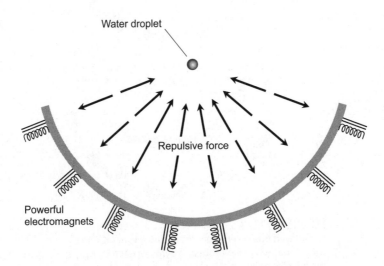

FIGURE 12-12 · Levitation can result from the action of powerful electromagnets against a diamagnetic substance, such as a droplet of water.

Superconductivity

At temperatures approaching *absolute zero* (about −273°C or −459°F), certain substances lose all of their electrical resistance in the practical sense, so they become *superconductive*. A small current in a superconducting wire loop can circulate without growing noticeably weaker for a long time indeed—provided that we don't connect a resistive load and thereby dissipate the current as heat. In theory, the current in a superconducting loop continues forever. In practice, the current gradually dies down, but it can remain significant for years.

Certain metals, such as aluminum and various alloys and compounds, exhibit superconductivity when they get cold enough. Other metals never get superconductive, no matter how low we drive the temperature. In a substance that can function as a superconductor, there exists a *critical temperature*, also called a *threshold temperature*, below which superconductivity takes place. Above that temperature, the substance behaves like a conventional electrical conductor, with small but measurable resistance. If we warm up a superconducting substance to the critical temperature, its resistance increases abruptly, and the superconductive properties vanish. Specially manufactured alloys called *cuprates* can superconduct at temperatures considerably above absolute zero.

Superconducting objects make magnetic levitation possible because they completely expel the magnetic flux, a phenomenon known as *perfect diamagnetism* or the *Meissner Effect* (after one of its discoverers, *Walter Meissner*, who first noticed it in the 1930s). A superconductor has permeability equal to 0 for all practical purposes. Superconductors repel strong magnetic fields with considerable force, a property that explains why superconductors have generated interest among engineers seeking to build massive machines that take advantage of magnetic levitation.

Rotation

The arrangement shown in Fig. 12-10A doesn't work because the upper magnet flips over with the tiniest instability. In practice, it can't stay centered for more than a few milliseconds before a puff of air or some other disturbance jostles it, and its south pole comes to rest against the north pole of one of the lower magnets.

There's an amazingly simple way to get an arrangement similar to that of Fig. 12-10A to produce levitation: spin the upper magnet at a rapid rate, and keep it spinning. We can attach a nonferromagnetic disk to the upper magnet to increase the gyroscopic effect produced by the rotation, as shown in Fig. 12-13.

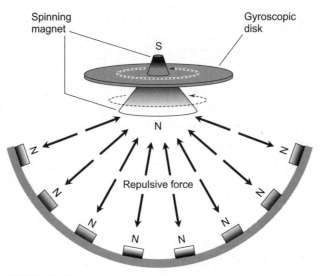

FIGURE 12-13 · A spinning magnet equipped with a gyroscopic disk can levitate above an array of fixed magnets. (The size of the upper magnet is exaggerated for clarity.)

(The size of the upper magnet is exaggerated for clarity.) The upper magnet acts as a spinning top that "balances" on the magnetic fields produced by itself and the lower array of magnets.

? Still Struggling

A system such as the one shown in Fig. 12-13 does not violate Earnshaw's Theorem. Remember that this theorem applies only to stationary magnets of constant intensity. Nevertheless, the system shown in Fig. 12-13 is temperamental. The spin rate must stay between certain rotational speed limits, and the spinning magnet must have precisely the correct shape. Unless we devise some means to keep the upper magnet spinning, air resistance will eventually slow it down to the point where the gyroscopic effect fails, and it flips over and "sticks" to one of the magnets in the lower array.

Oscillating Fields

A nonferromagnetic object that conducts electricity behaves as a diamagnetic material in the presence of an alternating (or oscillating) magnetic field. We can generate an oscillating magnetic field by applying high-frequency AC to a set of electromagnets whose core materials exhibit high permeability but low retentivity. A suitably shaped, rotating disk made of a material such as aluminum will levitate above an array of such electromagnets, as shown in Fig. 12-14.

This method of levitation works because of *eddy currents* that appear in the conducting disk. The eddy currents produce a secondary magnetic field that opposes the primary, oscillating field set up by the array of fixed AC electromagnets. This effect tends to expel the external magnetic flux from the disk, so that the disk behaves as an "antimagnet." The rotation provides a stabilizing, gyroscopic effect, so the disk doesn't flip over. The disk stays centered as long as we place it along the central axis of the array to begin with, and as long as no significant disturbance throws it off center.

Eventually the rotation rate of the disk will decay because of air resistance, and the system will becomes unstable. However, we can eliminate this "bugaboo" by placing the entire system in a vacuum. (We can do the same thing with the rotating-magnet system described in the previous section.) This scheme allows the apparatus to operate forever in theory, although in practice, it will eventually fail because of inevitable *kinetic energy loss*. We can get close to building a perpetual motion machine, but we can never actually do it!

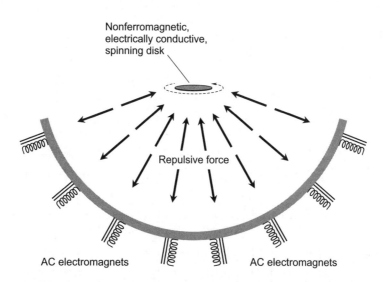

FIGURE 12-14 · A spinning disk, made of nonferromagnetic material that conducts electric currents, can levitate above a set of AC electromagnets.

The Maglev Train

For decades, high-speed rail transit has remained among the most talked-about applications of magnetic levitation. Some passenger trains for urban commuters employ this technology, which its proponents call *maglev*. The diamagnetic effects of superconductors are most often used for maglev systems.

The chief advantage of maglev trains over conventional trains lies in the fact that in a maglev train, friction occurs only between the moving carriages and the air, never between the carriages and the track. The train does not physically touch the track at any time, but hovers over the track as a result of the maglev effect. A gap of 2 to 3 centimeters (about an inch) prevails between the train and the track, which usually constitutes a *monorail* (single rail).

Maglev train cars can be supported by either of two geometries, as illustrated in Fig. 12-15. In either arrangement, vertical magnetic fields cause the train cars to remain suspended above the rail, and horizontal magnetic fields stabilize the cars so that they remain centered. Devices called *linear motors* provide forward or reverse train motion along the track.

Superconductor type maglev systems rely on numerous, powerful electromagnets embedded in the track to obtain the levitation. This arrangement raises the problem of shielding passengers from the strong magnetic fields. Magnetic shielding adds weight and expense to the construction of the train cars. Another potential problem is the ever-present danger of power failures. What will happen if a superconductor type maglev train traveling at high speed suddenly loses its support system, causing the cars to settle onto the track? Foul weather can also pose a hazard. What will happen to a maglev train in a sudden snowstorm, hailstorm, or windstorm if the track gets clogged with ice or the train cars get blown off center? If we really want to play the role of the "what-if troublemaker," we can ask what would happen if a massive earthquake took place.

One suggested way to mitigate the above-mentioned perils involves an alternative system that employs permanent magnets in the cars and wire loops in the track. The motion of the cars with respect to the track produces the levitation, in a manner similar to the way a rotating, conducting disk levitates above a set of fixed magnets. This system travels on sets of small wheels as it first gets going. Once it's moving at a few kilometers per hour, the currents in the loops become sufficient to set up magnetic fields that repel the permanent magnets in the train cars. As with the superconductor type maglev train, this system uses linear motors to achieve propulsion. If power fails, the cars coast to a stop, but they do not fall onto the track suddenly at high speed. They settle onto the

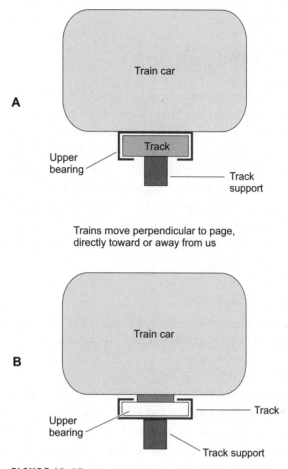

FIGURE 12-15 · Simplified cross-sectional diagrams of maglev train geometries. At A, an upper bearing, attached to the car, wraps around and levitates above a monorail track. At B, the upper bearing levitates inside a wrap-around track.

wheels once the speed drops below a few kilometers per hour. In the event of a storm that disrupts the track or the system balance, the entire train can be powered-down and it will, presumably, coast safely to a stop.

TIP *Maglev trains can travel at higher speeds than conventional trains do, but maglev trains cost a lot of money to build and maintain. Some engineers claim that the benefits of maglev trains fail to outweigh their high cost and technical challenges. Proponents cite the potential for lower noise, reduced commuting times, and the adaptability of maglev systems to the use of nonpolluting or renewable energy sources.*

? **Still Struggling**

You might wonder how a maglev train can negotiate a steep grade up a hill or down into a valley. Won't the train "slide" downgrade if no friction exists between the train and the track to provide braking action? The answer might surprise you! The linear motors used in maglev train systems can drive the cars up steeper grades than the locomotives for conventional trains could ever do. In addition, the linear motors can provide braking action by switching into reverse, and they can keep the train from "sliding" downgrade by operating against the force of gravity.

PROBLEM 12-9

How can we build a superconductor type maglev train to protect passengers from the strong fields that the electromagnets generate?

✔ SOLUTION

We can manufacture the train cars with thin external plating made from a ferromagnetic substance, such as steel, and we can take special care to ensure that the enclosure surrounds the passenger compartments sufficiently to keep the magnetic flux out. Unfortunately, steel has far greater mass per unit volume than aluminum, the other metal commonly used in rail-carriage construction. Aluminum does not have ferromagnetic properties, however, so it can't offer any protection against magnetic fields unless we force high, potentially dangerous electric currents through it.

QUIZ

This is an "open book" quiz. You may refer to the text in this chapter. You'll find the correct answers listed in the back of the book.

1. The repulsive force between a diamagnetic object and an electromagnet becomes significant if the
   A. permeability of the diamagnetic material is high enough.
   B. electromagnet can produce high enough flux density.
   C. temperature rises high enough.
   D. All of the above

2. Earnshaw's Theorem tells us that if we want to obtain magnetic levitation with a set of permanent magnets, then at least one of the magnets must
   A. be a monopole.
   B. be a bar magnet.
   C. have low retentivity.
   D. move or rotate.

3. The resistance of a superconducting material is theoretically
   A. zero.
   B. alternating.
   C. negative.
   D. infinite.

4. In the event of a power failure, a maglev train can be kept from "crashing" onto the track by means of
   A. gravitational force on the cars.
   B. small wheels on the cars.
   C. a vacuum between the cars and the track.
   D. linear motors in the track.

5. In a rod-shaped electromagnet operating from 100-Hz AC, the magnetic polarity at each end of the rod reverses once every
   A. 100th of a second.
   B. 50th of a second.
   C. 200th of a second.
   D. 500th of a second.

6. We wrap a coil of wire around a ferromagnetic rod. We find that we can make the magnetic flux density as great as 800 G inside the rod by driving DC through the coil. Further increases in coil current cause no further increase in the flux density inside the rod. When we remove the current from the coil, the flux density inside the rod drops to 80 G. What's the retentivity of the core material?
   A. 0.01
   B. 0.1
   C. 10
   D. 100

7. What's the permeability of the core material described in Question 6?

   A. 100
   B. 800
   C. 64,000
   D. We need more information to answer this question.

8. When the DC in the wire coil described in Question 6 reaches or exceeds the level necessary to produce a flux density of 800 G in the core, then the core material operates in a state of

   A. saturation.
   B. remanence.
   C. flux dilation.
   D. diamagnetism.

9. Which of the following characteristics represents an advantage of a magnetic disk over a magnetic tape for data storage and retrieval?

   A. Higher read/write speed
   B. Longer data life
   C. Immunity to external magnetic fields
   D. Better sensitivity

10. The data on a magnetic recording medium can be adversely affected by

   A. bright light.
   B. X rays.
   C. high temperatures.
   D. All of the above

Test: Part III

Do not refer to the text when taking this test. You may draw diagrams or use a calculator if necessary. A good score is at least 38 correct. You'll find the answers listed in the back of the book. Have a friend check your score the first time, so you won't memorize the answers if you want to take the test again.

1. If we live in Canada, we can expect to see the aurora borealis ("northern lights") shortly after
   A. a thunderstorm.
   B. a hurricane.
   C. an earthquake.
   D. a volcanic eruption.
   E. a solar flare.

2. When we wind a coil of wire around a magnetic compass and then drive DC through the wire, we can expect that the compass needle will deflect because of
   A. wire resistance.
   B. voltage interaction.
   C. galvanism.
   D. geomagnetism.
   E. magnetic inclination.

3. Webers per square meter is an alternative way to express the
   A. ampere.
   B. volt.
   C. watt.
   D. tesla.
   E. kilowatt-hour.

4. We can *never* achieve magnetic levitation with a set of
   A. electromagnets of any kind.
   B. rotating electromagnets.
   C. stationary, nonrotating permanent magnets.
   D. permanent magnets of any kind.
   E. any components other than superconductors.

5. Suppose that we find a sample of material and discover that it concentrates magnetic lines of flux to some extent. However, we can't make it into a permanent magnet, no matter how hard we try. By definition, this type of material is
   A. ferromagnetic.
   B. paramagnetic.
   C. diamagnetic.
   D. quasimagnetic.
   E. antimagnetic.

6. If we double the frequency of an electromagnetic (EM) wave traveling through free space, the propagation speed
   A. becomes 1/4 as great.
   B. becomes 1/2 as great.
   C. does not change.
   D. doubles.
   E. quadruples.

7. Imagine that we have a DC-carrying loop of wire surrounded by, and containing, a
    vacuum. Then we place a sample of *diamagnetic* material somewhere inside the
    loop. If we keep the same steady DC flowing in the loop, we can have *complete con-
    fidence* that one of the following things will happen inside the material. Which one?

    A. The magnetic flux density will decrease.
    B. The magnetic flux density will increase.
    C. The magnetomotive force will decrease.
    D. The magnetomotive force will increase.
    E. The magnetic retentivity will increase.

8. Imagine that we have a DC-carrying loop of wire surrounded by, and containing, a
    vacuum. Then we place a sample of *ferromagnetic* material somewhere inside the
    loop. If we keep the same steady DC flowing in the loop, we can have *complete con-
    fidence* that one of the following things will happen inside the material. Which one?

    A. The magnetic flux density will decrease.
    B. The magnetic flux density will increase.
    C. The magnetomotive force will decrease.
    D. The magnetomotive force will increase.
    E. The magnetic retentivity will increase.

9. Which of the following scenarios will *always* produce an attractive magnetic force?

    A. The north pole of a permanent magnet brought near a piece of iron
    B. The south pole of a permanent magnet brought near a piece of glass
    C. The north pole of a permanent magnet brought near the north pole of another
        permanent magnet
    D. Two ferromagnetic objects brought near each other
    E. Any of the above

10. Which of the following types of energy *never* occurs as an EM field?

    A. Visible-light rays
    B. X rays
    C. Gamma rays
    D. Sound waves
    E. Radio waves

11. In Fig. Test III-1, object X could be

    A. a rod-shaped DC electromagnet.
    B. a rod-shaped permanent magnet.
    C. a rod-shaped AC electromagnet.
    D. a cylindrical wire coil carrying DC.
    E. Any of the above

12. In Fig. Test III-1, which end of the rectangle represents a magnetic south pole?

    A. The left-hand end
    B. The right-hand end
    C. Neither end
    D. Both ends
    E. We need more information to answer this question.

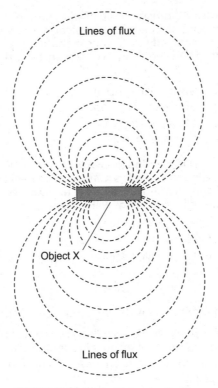

Lines of flux

Object X

Lines of flux

FIGURE TEST III-1 · Illustration for Part
III Test Questions 11 and 12.

13. **The magnetic permeability of a vacuum is**
    A. negative.
    B. equal to 0.
    C. equal to 1.
    D. greater than 1.
    E. theoretically infinite.

14. **Suppose that we wrap a coil of wire around a rod-shaped ferromagnetic core. We find that we can make the magnetic flux density as great as 640 G inside the core by driving DC through the coil until the core material saturates. When we remove the current from the coil, the flux density inside the core drops to 32 G. What's the permeability of the core material?**
    A. 5.0%
    B. 0.25%
    C. 12.5%
    D. 22%
    E. We need more information to answer this question.

15. **In Fig. Test III-2, object X could be**

A. a bar magnet.
B. an electrically charged particle.
C. a DC electromagnet.
D. an AC electromagnet.
E. a wire loop carrying DC.

16. **If you indefinitely increase the frequency of the AC applied to an electromagnet, the intensity of the magnetic field will start to decline sooner or later. Why?**

A. The coil conducts current better and better as the frequency increases, behaving as a superconductor at extremely high AC frequencies.
B. Retentivity interferes, to an increasing extent, with the ability of the magnetic flux from the coil to magnetize the core as the frequency goes up.
C. The magnetic field in the core opposes the magnetic field produced by the coil to a greater extent as the frequency increases.
D. The effective voltage between the ends of the coil decreases, approaching zero as the frequency gets extremely high.
E. All of the above

17. **A magnetic tape drive can store computer files and programs for long periods without the need for a source of power to maintain the integrity of the data. For this reason, magnetic tape is, by definition,**

A. a reversible storage medium.
B. a diamagnetic storage medium.
C. a nonvolatile storage medium.
D. a read-only storage medium.
E. a flash-memory storage medium.

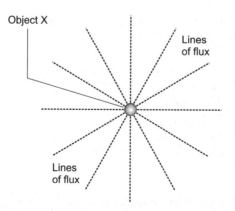

FIGURE TEST III-2 · Illustration for Part III
Test Question 15.

18. If we double the wavelength of an electromagnetic (EM) wave traveling through free space, the frequency
    A. becomes 1/4 as great.
    B. becomes 1/2 as great.
    C. does not change.
    D. doubles.
    E. quadruples.

19. Which of the following statements holds true for EM waves in free space when we express the wavelength in meters, the frequency in hertz, and the propagation speed in meters per second?
    A. The propagation speed equals the frequency times the wavelength.
    B. The frequency equals the wavelength times the propagation speed.
    C. The wavelength equals the frequency times the propagation speed.
    D. The wavelength and the frequency vary in direct proportion to each other.
    E. None of the above

20. The ideal core material for a powerful AC electromagnet has
    A. high retentivity and high permeability.
    B. high retentivity and low permeability.
    C. low retentivity and high permeability.
    D. low retentivity and low permeability.
    E. infinite retentivity and zero permeability.

21. You should never use an automotive battery in an attempt to make a "super-powerful" DC electromagnet because the
    A. electromagnet could draw excessive current from the battery, causing battery acid to boil out.
    B. battery voltage isn't high enough.
    C. battery voltage is too high.
    D. battery can't deliver sufficient current to make an electromagnet work.
    E. battery produces AC, not DC.

22. Which of the following characteristics of a home-entertainment system increases its vulnerability to electromagnetic interference (EMI)?
    A. The use of long connecting cables
    B. The use of unshielded connecting cables
    C. The absence of an electrical ground
    D. The use of numerous peripheral devices
    E. All of the above

23. Which of the following forms of radiation has the highest frequency?
    A. Radio "short waves"
    B. X rays
    C. Visible green light

    D. Radio microwaves

    E. Visible violet light

24. **Which of the following forms of radiation has the lowest frequency?**

    A. Radio "short waves"

    B. X rays

    C. Visible green light

    D. Radio microwaves

    E. Visible violet light

25. **Figure Test III-3 is a functional diagram of a**

    A. fluxgate magnetometer.

    B. dynamic transducer.

    C. maglev locator.

    D. retentivity detector.

    E. relay navigation system.

26. **Suppose that we encounter a stepper motor carrying its rated current, but whose shaft remains in a fixed position (nonrotating). If we grab the shaft and try to turn it, what should we expect?**

    A. The shaft will freely and easily rotate.

    B. We'll have trouble turning the shaft; it will "try" to stay in place.

    C. The shaft will begin to rotate in defined, timed increments (steps).

    D. The motor will overheat, and might burn out.

    E. The motor will overload its power supply, blowing the fuse.

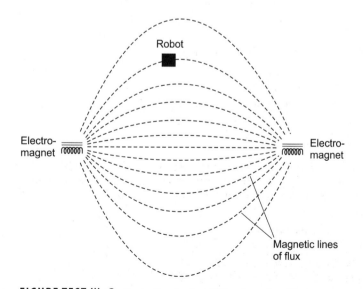

FIGURE TEST III - 3 · Illustration for Part III Test Question 25.

27. Suppose that a galvanometer coil carries 100 mA of DC, causing the compass needle to deflect 50° to the west of geomagnetic north. What will the needle do if we increase the current to 150 mA without changing its direction?

    A. The needle will deflect to the west of geomagnetic north by 50°.
    B. The needle will deflect to the west of geomagnetic north by some angle smaller than 50°.
    C. The needle will deflect to the west of geomagnetic north by some angle greater than 50°.
    D. The needle will deflect to the east of geomagnetic north by some angle greater than 50°.
    E. The needle will deflect to the east of geomagnetic north by some angle smaller than 50°.

28. Fill in the blank to make the following statement true: "In a cylindrical DC electromagnet with a ferromagnetic core, the flux density inside the core material varies in *inverse proportion* relative to the _____, assuming all other factors remain constant."

    A. permeability of the core material
    B. current in the coil
    C. number of turns in the coil
    D. end-to-end length of the coil
    E. More than one of the above

29. In an electromagnetic cathode-ray tube (CRT), the electron gun

    A. deflects the electron beam.
    B. speeds up the electron beam.
    C. slows down the electron beam.
    D. causes the electron beam to sweep across the screen.
    E. generates the electron beam.

30. In an electromagnetic CRT, the anodes

    A. deflect the electron beam.
    B. speed up the electron beam.
    C. slow down the electron beam.
    D. cause the electron beam to sweep across the screen.
    E. generate the electron beam.

31. If we cut in half the *number of turns* in a multiple-turn circular wire loop carrying DC but we don't change the current or the loop diameter, the magnetomotive force at a fixed point in or near the loop

    A. increases by a factor of 4.
    B. increases by a factor of 2.
    C. does not change.
    D. decreases to 1/2 of its previous value.
    E. decreases to 1/4 of its previous value.

32. If we cut in half the *diameter* of a multiple-turn circular wire loop carrying DC but we don't change the current or the number of turns, the magnetomotive force at a fixed point in or near the loop

    A. increases by a factor of 4.
    B. increases by a factor of 2.
    C. does not change.
    D. decreases to 1/2 of its previous value.
    E. decreases to 1/4 of its previous value.

33. Fill in the blank to make the following statement true: "In a cylindrical DC electromagnet with a ferromagnetic core, the flux density inside the core material varies in *direct proportion* relative to the _____ , assuming all other factors remain constant."

    A. permeability of the core
    B. current in the coil
    C. number of turns in the coil
    D. length of the coil
    E. More than one of the above

34. Suppose that we wrap a coil of wire around a donut-shaped ferromagnetic core. We find that we can make the magnetic flux density as great as 500 G inside the core by driving DC through the coil until the core material saturates. When we remove the current from the coil, the flux density inside the core drops to 100 G. What's the retentivity of the core material?

    A. 4.0%
    B. 20%
    C. 25%
    D. 45%
    E. We need more information to answer this question.

35. You should never plug a coil into a utility outlet in an attempt to make a "super-powerful" AC electromagnet because the

    A. electromagnet could produce dangerous radiation.
    B. utility voltage reverses its polarity too fast.
    C. utility voltage doesn't reverse its polarity fast enough.
    D. high current demand could overload your household electrical wiring.
    E. coil and core combination will, in effect, convert the AC to DC.

36. In order to produce an EM field, electric charge carriers must

    A. exist in pairs.
    B. move at steady speed and constant direction.
    C. accelerate or decelerate.
    D. form dipoles.
    E. All of the above

37. The *magnetic flux density* at the center of a multiple-turn, circular wire loop depends on the

    A. current in the loop.
    B. radius of the loop.
    C. number of turns in the loop.
    D. permeability of the medium inside the loop.
    E. All of the above

38. The *magnetomotive force* at the center of a multiple-turn, circular wire loop depends on the

    A. current in the loop.
    B. circumference of the loop.
    C. gauge of the wire used to wind the loop.
    D. permeability of the medium inside the loop.
    E. None of the above

39. Which of the following free-space wavelengths is typical of extremely-low-frequency (ELF) fields?

    A. 50 nm
    B. 500 nm
    C. 500 mm
    D. 50 cm
    E. None of the above

40. Distortion occurs in the magnetic flux field in space around the earth as a result of

    A. magnetic declination.
    B. magnetic inclination.
    C. the solar wind.
    D. the earth's orbital motion around the sun.
    E. All of the above

41. Which of the following statements can we correctly make concerning the ELF fields from CRT computer monitors?

    A. Some people believe that they might pose a health hazard to humans.
    B. They can make any material object in their vicinity become radioactive.
    C. They can cause severe burns in exposed living tissue.
    D. They can cause short-term partial blindness.
    E. All of the above

42. A device that uses a coil and magnet to convert sound waves into electric current or vice-versa constitutes an example of

    A. an AC motor.
    B. a fluxgate magnetometer.
    C. a dynamic transducer.
    D. a selsyn.
    E. a synchro.

43. **We can express magnetomotive force in terms of**
    A. volts per square meter.
    B. gilberts.
    C. watts per square meter.
    D. teslas.
    E. joules per watt.

44. **Scientists sometimes use the Ångström to express**
    A. magnetic flux density.
    B. electric flux density.
    C. magnetomotive force.
    D. EM wavelength.
    E. electric charge gradient.

45. **The geomagnetic lines of flux converge at**
    A. the north geomagnetic pole.
    B. the north geographic pole.
    C. the geomagnetic equator.
    D. the geographic equator.
    E. None of the above

46. **The magnetic lines of flux around a long, straight, DC-carrying wire diverge from, or converge toward,**
    A. all points "infinitely far" from the wire.
    B. the ends of the wire.
    C. all points on the wire.
    D. multiple points evenly spaced along the wire.
    E. None of the above

47. **Figure Test III-4 illustrates a scheme that has been successfully implemented to obtain magnetic levitation. Why does it work?**
    A. The conductive properties of the disk prevent magnetic fields from penetrating or influencing it in any way. Actually, the disk can levitate, but it occurs because of the electric fields produced by the voltages in the electromagnets. It has nothing to do with magnetism.
    B. Eddy currents appear in the conducting disk, producing a field that opposes the oscillating field set up by the array of fixed AC electromagnets. In addition, the rotation stabilizes the disk.
    C. The disk becomes diamagnetic, so it repels magnets of all kinds; the strong electromagnets produce force powerful enough to suspend the disk.
    D. The disk becomes superconductive, so it turns into an incredibly powerful permanent magnet that repels the AC electromagnets.
    E. The premise of the question is false. According to electromagnetic theory, it can't work, and no one has ever made it work.

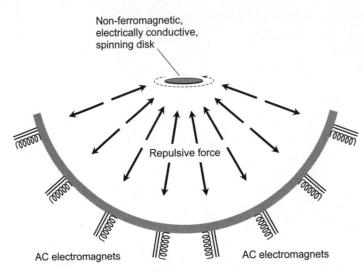

Non-ferromagnetic, electrically conductive, spinning disk

Repulsive force

AC electromagnets   AC electromagnets

FIGURE TEST III-4 · Illustration for Part III Test Questions 47 and 48.

48. **If you answered E to Question 47, why can't the magnetic-levitation arrangement shown in Figure Test III-4 work?**

    A. Because of the constraints imposed by Earnshaw's theorem.
    B. Because the disk can't magnetize sufficiently.
    C. Because we can't get the disk to rotate fast enough.
    D. Because we can't make AC electromagnets work at high enough frequencies.
    E. Answer E is wrong for Question 47. The arrangement can and does work.

49. **We can express the overall quantity of a magnetic field in terms of**

    A. teslas.
    B. gauss.
    C. maxwells.
    D. watts per meter.
    E. volts per square meter.

50. **As an EM field travels through free space, the electric lines of flux and the direction of propagation at any given point**

    A. run parallel to each other.
    B. run perpendicular to each other.
    C. rotate in opposite senses.
    D. rotate in the same sense.
    E. coincide.

Final Exam

Do not refer to the text when taking this test. You may draw diagrams or use a calculator if necessary. A good score is at least 75 correct. You'll find the answers listed in the back of the book. Have a friend check your score the first time, so you won't memorize the answers if you want to take the test again.

1. **The voltage produced by an electrochemical battery depends on**
   A. the total combined surface area of the cell plates.
   B. the chemical composition of the cells.
   C. the total combined volume of all the cells.
   D. the total combined mass of all the cells.
   E. All of the above

2. **The maximum power output that an electrochemical battery can deliver depends on**
   A. the total combined surface area of the cell plates.
   B. the chemical composition of the cells.
   C. the total combined volume of all the cells.
   D. the total combined mass of all the cells.
   E. All of the above

3. **We can have *absolute confidence* that an atom carries no net electric charge if we know that it contains**
   A. the same number of electrons as neutrons.
   B. the same number of electrons as protons.
   C. the same number of protons as neutrons.
   D. more electrons than protons.
   E. more protons than electrons.

4. **We can have *absolute confidence* that an atom carries a net positive electric charge if we know that it contains**
   A. the same number of electrons as neutrons.
   B. the same number of electrons as protons.
   C. the same number of protons as neutrons.
   D. more electrons than protons.
   E. more protons than electrons.

5. **Figure Exam-1 shows a meter connected through a switch to an electrical system. Based on the appearance of this block diagram, what type of meter does the circle with the arrow most likely represent?**
   A. An ohmmeter
   B. A voltmeter
   C. A milliammeter
   D. A galvanometer
   E. A fluxgate magnetometer

6. **In the scenario of Fig. Exam-1, the meter measures the electricity at the input to the**
   A. computer.
   B. hi-fi set.
   C. motor.

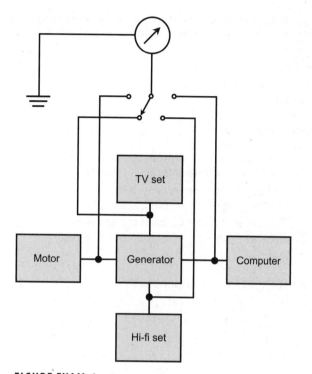

FIGURE EXAM-1 · Illustration for Final Exam Questions 5 and 6.

   D. TV set.
   E. More than one of the above

7. **What's the period of an AC wave whose frequency equals 33.333 kHz?**
   A. 30.000 μs
   B. 300.00 μs
   C. 30.000 ns
   D. 300.00 ns
   E. It depends on the waveform.

8. **Imagine that we charge up a secondary cell to its full extent, and then connect it to a load that draws a continuous current of 500 mA. This amount of current flows steadily through the load for exactly 12 hours, and then the cell suddenly "dies." The storage capacity of this cell was evidently**
   A. 6.0 Ah.
   B. 4.0 Ah.
   C. 3.6 Ah.
   D. 2.4 Ah.
   E. 1.8 Ah.

9. Which of the following characteristics should we expect to observe if we use a voltage-doubler power supply in an application where the load draws a large, variable current?

A. Poor linearity
B. Complete lack of ripple
C. Unusual sensitivity to transients
D. Lack of forward breakover
E. Poor voltage regulation

10. Figure Exam-2 shows five 12-V light bulbs connected to a 12-V battery along with some switches and potentiometers. Component X

A. affects the brilliance of the uppermost bulb only.
B. affects the brilliance of the lowermost bulb only.
C. affects the brilliance of all the bulbs at the same time.
D. prevents the other potentiometers from burning out.
E. None of the above

11. In the circuit of Fig. Exam-2, component Y

A. switches all the bulbs off or on at once.
B. switches the second bulb from the top off or on, but has no effect on any other bulb.

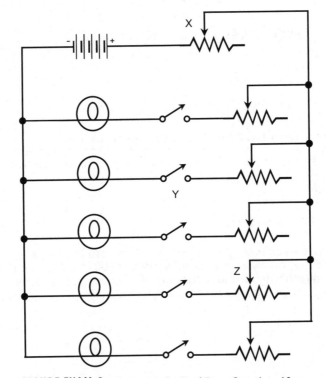

FIGURE EXAM-2 · Illustration for Final Exam Questions 10 through 12.

C. switches all of the bulbs off or on at the same time, except for the second bulb from the top, which remains glowing no matter what.

D. allows the adjacent potentiometer to work when open, but keeps that potentiometer from having an effect when closed.

E. allows all of the potentiometers to work when open, but keeps any of them from having an effect when closed.

12. **In the circuit of Fig. Exam-2, component Z**

A. affects the brilliance of the uppermost bulb only.
B. affects the brilliance of the lowermost bulb only.
C. affects the brilliance of all the bulbs at the same time.
D. affects the brilliance of every bulb except the one second from the bottom.
E. None of the above

13. **Suppose that the AC input voltage to a step-up transformer equals 120 V RMS, and constitutes a pure sine wave with no DC component. The primary-to-secondary turns ratio equals precisely 1:2. What's the peak-to-peak voltage across the secondary?**

A. 679 V pk-pk
B. 339 V pk-pk
C. 170 V pk-pk
D. 84.9 V pk-pk
E. 42.4 V pk-pk

14. **In a three-wire AC system outlet designed for common 117-V household use, the electrical ground should always connect directly to**

A. both rectangular slots and the round or D-shaped hole.
B. the round or D-shaped hole.
C. the narrower of the two rectangular slots.
D. both rectangular slots, but not the round or D-shaped hole.
E. neither of the two rectangular slots nor the round or D-shaped hole.

15. **When the voltage across a component increases but the resistance does not change, the current through that component varies according to one of the curves shown in Fig. Exam-3. Which one?**

A. Curve A
B. Curve B
C. Curve C
D. Curve D
E. Curve E

16. **In theory, when exposed to bright sunlight, a battery comprising 81 individual 600-mV silicon photovoltaic (PV) cells in a 9-by-9 series-parallel matrix should produce**

A. 600 mV DC.
B. 1.80 V DC.
C. 3.60 V DC.
D. 5.40 V DC.
E. 48.6 V DC.

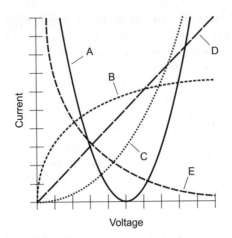

FIGURE EXAM-3 · Illustration for Final
Exam Question 15.

17. Fill in the blank to make the following statement true: "As the total electrostatic charge quantity in two objects increases, and if the distance between the centers of the objects doesn't change, the electrostatic force between the objects increases in _____ the total charge quantity."
    A. direct proportion to the square of
    B. direct proportion to
    C. direct proportion to the square root of
    D. inverse proportion to
    E. inverse proportion to the square of

18. As the separation between the centers of two charged objects increases, and if their total combined electrostatic charge quantity remains constant, the electrostatic force between the objects
    A. decreases in direct proportion to the square root of the distance between their centers.
    B. decreases in direct proportion to the distance between their centers.
    C. decreases in direct proportion to the square of the distance between their centers.
    D. increases in direct proportion to the distance between their centers.
    E. increases in direct proportion to the square of the distance between their centers.

19. Figure Exam-4 shows a DC source, a resistor, and an ammeter connected in series. Let E represent the DC source voltage. Let I represent the current in the circuit. Let R represent the value of the resistor. According to Ohm's Law, if $E = 25.0$ V and $R = 100$ ohms, then
    A. $I = 250$ mA.
    B. $I = 2.500$ A.
    C. $I = 4.00$ A.
    D. $I = 6.25$ A.
    E. $I = 160$ mA.

20. In the scenario of Fig. Exam-4, suppose that $I = 800$ µA and $R = 150$ k. In that case,

    A. $E = 188$ mV.
    B. $E = 5.33$ V.
    C. $E = 120$ V.
    D. $E = 427$ V.
    E. $E = 6.94$ V.

21. Suppose that, in the circuit of Fig. Exam-4, we double the resistance and cut the DC source voltage in half. In that case, the current

    A. increases by a factor of 4.
    B. doubles.
    C. stays the same.
    D. drops to half of its previous value.
    E. drops to 1/4 of its previous value.

22. Suppose that, in the circuit of Fig. Exam-4, we want to make changes that result in the current increasing by a factor of 100. To accomplish this task, we can

    A. multiply the DC source voltage by 100 and leave the resistance the same.
    B. decrease the resistance to 1/100 of its previous value and leave the DC source voltage the same.
    C. decrease the resistance to 1/10 of its previous value and increase the DC source voltage by a factor of 10.
    D. decrease the resistance to 1/5 of its previous value and increase the DC source voltage by a factor of 20.
    E. Do any of the above

23. In the scenario of Fig. Exam-4, suppose that $I = 63.2$ mA and $E = 15.73$ V. In that case,

    A. $R = 401$ ohms.
    B. $R = 249$ ohms.
    C. $R = 1.61$ k.
    D. $R = 6.19$ M.
    E. $R = 0.994$ ohms.

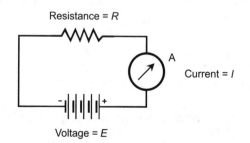

FIGURE EXAM-4 · Illustration for Final Exam Questions 19 through 23.

24. Figure Exam-5 shows a circuit containing a source of variable DC voltage, a potentiometer, an ammeter to measure the current through the potentiometer, a switch that allows us to place the potentiometer in or out of the circuit at will, a timer to measure the length of time that the potentiometer carries current while the switch remains closed, and a voltmeter to measure the EMF supplied by the voltage source. Suppose that we connect a 12.0-V battery as the source. We close the switch and read the ammeter, which tells us that $I = 800$ mA. Based on this information, we know that the potentiometer dissipates

    A. 5.56 W.
    B. 7.68 W.
    C. 9.60 W.
    D. 15.0 W.
    E. 18.8 W.

25. Refer again to Fig. Exam-5. Suppose that we leave the switch closed in the system of Question 24, and then we cut the DC source voltage in half and simultaneously increase the potentiometer's resistance to four times its previous value. The power that the potentiometer dissipates will

    A. remain unchanged.
    B. diminish to half of its previous value.
    C. diminish to 1/4 of its previous value.
    D. diminish to 1/8 of its previous value.
    E. diminish to 1/16 of its previous value.

26. Suppose that we connect a 24-V battery as the DC source in the system of Fig. Exam-5. We close the switch and read the ammeter, which tells us that $I = 2.5$ A. We leave the switch closed for 10 min. Based on this information, we know that during this period of time, the potentiometer dissipates

    A. 10 Wh.
    B. 25 Wh.
    C. 20 Wh.
    D. 30 Wh.
    E. 60 Wh.

27. Suppose that we repeat the experiment described in Question 26, but with a 12-V battery instead of a 24-V battery as the DC source. At the same time, we cut the resistance of the potentiometer in half. In 10 min, the potentiometer will dissipate

    A. 1/4 as much energy as it did before.
    B. half as much energy as it did before.
    C. the same amount of energy as it did before.
    D. twice as much energy as it did before.
    E. four times as much energy as it did before.

28. Coaxial cable works well for radio-frequency (RF) signal transmission because it has

    A. four conductors.
    B. stranded conductors.
    C. electromagnetic (EM) shielding.

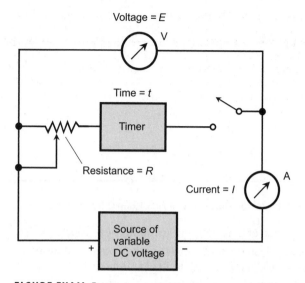

FIGURE EXAM-5 • Illustration for Final Exam Questions 24 through 27.

D. no dielectric barrier.

E. high resistance per unit length.

29. **An automotive battery comprises**

A. rechargeable cells.

B. zinc-carbon cells.

C. alkaline cells.

D. mercury cells.

E. photovoltaic cells.

30. **Figure Exam-6 on page 328 shows four batteries called W, X, Y, and Z connected in series, with voltages as indicated. In theory, the voltage E equals**

A. 36 V.

B. 30 V.

C. 27 V.

D. 24 V.

E. 18 V.

31. **In theory, if we reverse the polarity of battery Y in the scenario of Fig. Exam-6 but leave all the other batteries connected the same way, the voltage E becomes**

A. 6.0 V.

B. 7.5 V.

C. 9.0 V.

D. 12 V.

E. 15 V.

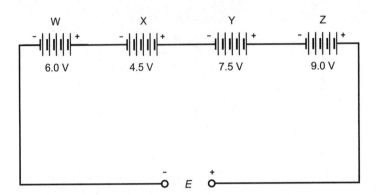

FIGURE EXAM-6 · Illustration for Final Exam Questions 30 and 31.

32. **The DC resistance of a fixed length of solid, pure copper wire**
    A. varies inversely in proportion to the square of its diameter.
    B. varies inversely in proportion to its diameter.
    C. does not depend on its diameter.
    D. varies directly in proportion to its diameter.
    E. varies directly in proportion to the square of its diameter.

33. **What's the current through the resistor marked X in the circuit of Fig. Exam-7?**
    A. 3 A
    B. 5 A
    C. 8 A
    D. 10 A
    E. We need more information to calculate it.

34. **What's the value of the resistor marked X in the circuit of Fig. Exam-7?**
    A. 4 ohms
    B. 6 ohms
    C. 8 ohms
    D. 11 ohms
    E. We need more information to calculate it.

35. **What's the voltage across the resistor marked Y in the circuit of Fig. Exam-8?**
    A. 3 V
    B. 6 V
    C. 9 V
    D. 12 V
    E. We need more information to calculate it.

36. **Suppose that the series combination of resistors draws 3 A from the battery in the circuit of Fig. Exam-8. What's the value of the resistor marked Y?**
    A. 4 ohms
    B. 3 ohms

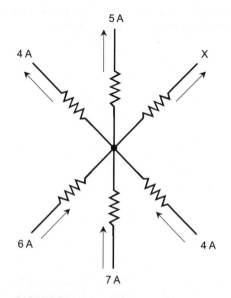

FIGURE EXAM-7 · Illustration for Final
Exam Questions 33 and 34.

    C. 2 ohms
    D. 1 ohm
    E. We need more information to calculate it.

**37. In a coaxial cable designed to keep the center conductor running exactly along
the central axis of the shield, engineers commonly employ**

    A. electromagnets.
    B. foamed or solid dielectric material.
    C. carbon granules.
    D. ground aluminum, copper, or steel.
    E. a vacuum.

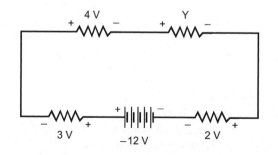

FIGURE EXAM-8 · Illustration for Final Exam
Questions 35 and 36.

38. If we connect a 47-ohm resistor, a 68-ohm resistor, and a 100-ohm resistor in *parallel*, the total resistance equals approximately

    A. 22 ohms.

    B. 33 ohms.

    C. 68 ohms.

    D. 72 ohms.

    E. 215 ohms.

39. If we connect a 47-ohm resistor, a 68-ohm resistor, and a 100-ohm resistor in *series*, the total resistance equals approximately

    A. 22 ohms.

    B. 33 ohms.

    C. 68 ohms.

    D. 72 ohms.

    E. 215 ohms.

40. Imagine four samples of solid wire made from four different metals: pure steel, pure aluminum, pure copper, and pure silver. All four samples have circular cross sections, and all four samples have the same American Wire Gauge (AWG) specification. Which of the four samples has the lowest DC resistance per unit length at room temperature?

    A. The steel sample

    B. The aluminum sample

    C. The copper sample

    D. The silver sample

    E. We need more information to answer this question.

41. As we draw a schematic diagram, suppose that we must allow two lines (representing wires) to cross over each other. We want to indicate that the two wires connect electrically at the point where they cross in the diagram. How can we *best* show this connection?

    A. Simply let the wires cross; don't add anything at the point of intersection.

    B. Place a small jog or "bump" in one of the lines at the place where the two lines cross.

    C. Place a heavy black dot at the point where the wires cross.

    D. Show one of the wires as two separate lines, each of which comes to the other line at a distinct point; place heavy black dots at both points.

    E. Place an open circle at the point where the lines cross.

42. We would most likely use clip leads to

    A. connect a battery to a test lamp.

    B. transfer serial data between two computers.

    C. transfer parallel data between two computers.

    D. connect a television set to a community antenna system.

    E. connect the utility AC wiring to a laundry machine.

43. **We should not expect a permanent magnet to "stick" to any part of**
    A. another permanent magnet.
    B. a glass window pane.
    C. a length of steel wire.
    D. a block of iron.
    E. Any of the above

44. **Suppose that an irregular AC wave has a positive peak voltage of +4.80 V pk+ and a negative peak voltage of −2.00 V pk−. What's the peak-to-peak voltage?**
    A. 1.40 V pk-pk
    B. 2.00 V pk-pk
    C. 4.80 V pk-pk
    D. 6.80 V pk-pk
    E. We need more information to answer this question.

45. **What's the root-mean-square (RMS) voltage of the irregular AC wave described in Question 44?**
    A. 0.495 V RMS
    B. 0.707 V RMS
    C. 1.70 V RMS
    D. 2.40 V RMS
    E. We need more information to answer this question.

46. **What's the frequency of an AC wave whose period equals 250 ns?**
    A. 800 kHz
    B. 1.25 MHz
    C. 2.50 MHz
    D. 4.00 MHz
    E. 8.00 MHz

47. **Imagine that someone shoots a steady stream of electrons in a straight line directly at your face. (It's not intense enough to harm you.) If you could see the magnetic flux caused by this electron beam, it would look like**
    A. straight lines coming straight toward you.
    B. straight lines radiating outward from a central point.
    C. concentric rings revolving clockwise around a central point.
    D. straight lines converging inward toward a central point.
    E. concentric rings revolving counterclockwise around a central point.

48. **The number of gilberts that a DC-carrying coil produces depends on**
    A. the current only.
    B. the number of turns only.
    C. the coil diameter, the number of turns, and the current.
    D. the current and the number of turns only.
    E. the current, the coil diameter, the number of turns, and the type of material inside the coil.

49. You can sometimes get rid of memory drain in an old nickel-cadmium (NICAD) battery by
   A. quick-charging it with an exceptionally high current.
   B. charging it with the polarity reversed.
   C. repeatedly discharging and recharging it.
   D. shorting it out and then recharging it once.
   E. connecting it to a source of low-voltage AC for a few hours.

50. Suppose that a certain load constitutes a pure resistance of 600 ohms. We measure an AC current through it as 150 mA RMS. How much power does the load dissipate?
   A. 13.5 W
   B. 6.25 W
   C. 27.0 W
   D. 90.0 W
   E. 400 W

51. Suppose that a certain load constitutes a pure resistance of 10 ohms. We measure an AC voltage across it as 20 V RMS. How much energy does the load dissipate in 120 minutes?
   A. 40 Wh
   B. 80 Wh
   C. 200 Wh
   D. 283 Wh
   E. 400 Wh

52. If we cut the resistance to 5.0 ohms in the scenario of Question 51 but don't change the applied AC voltage, what happens to the amount of energy that the load dissipates in 120 min?
   A. It drops to 1/4 of its previous value.
   B. It drops to half of its previous value.
   C. It does not change.
   D. It doubles.
   E. It quadruples.

53. Which of the following materials makes a good dielectric medium?
   A. Iron
   B. Copper
   C. Glass
   D. Aluminum
   E. Silver

54. A transistor battery that comes in a small, domino-shaped package is equipped with clamp-on terminals, provides approximately 9 V DC, and consists of several zinc-carbon or alkaline cells in series. How many cells are in the battery?
   A. Two
   B. Three

  C. Four
  D. Six
  E. Eight

55. **The conversion efficiency of a PV cell equals the ratio of**
  A. output current to output voltage.
  B. output voltage to output current.
  C. incident light power to useful output power.
  D. useful output power to incident light power.
  E. output power to internal resistance.

56. **Figure Exam-9 shows a waveform as it might appear on a laboratory oscillo-scope. Each vertical division represents 100 mV, and each horizontal division represents 50 ns. What's the positive peak voltage?**
  A. +0.200 V pk+
  B. +0.400 V pk+
  C. +0.800 V pk+
  D. +1.00 V pk+
  E. We need more information to say.

57. **What's the negative peak voltage of the wave shown in Fig. Exam-9?**
  A. −0.500 V pk−
  B. −0.400 V pk−
  C. −0.200 V pk−
  D. −0.100 V pk−
  E. We need more information to say.

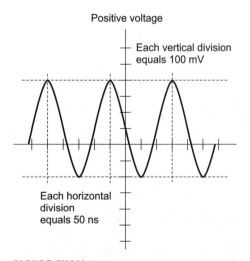

Positive voltage

Each vertical division equals 100 mV

Each horizontal division equals 50 ns

FIGURE EXAM-9 • Illustration for Final Exam Questions 56 through 61.

58. What's the peak-to-peak voltage of the wave shown in Fig. Exam-9?
    A. 600 mV pk-pk
    B. 1000 mV pk-pk
    C. 1200 mV pk-pk
    D. 1600 mV pk-pk
    E. We need more information to say.

59. Assuming that the waveform in Fig. Exam-9 precisely follows the shape of a sinusoid, what's its average voltage?
    A. +600 mV avg
    B. +500 mV avg
    C. +200 mV avg
    D. +100 mV avg
    E. 0.00 mV avg

60. What's the period of the waveform shown in Fig. Exam-9?
    A. 400 ns
    B. 200 ns
    C. 100 ns
    D. 40 ns
    E. 20 ns

61. What's the frequency of the waveform shown in Fig. Exam-9?
    A. 0.40 MHz
    B. 0.80 MHz
    C. 1.2 MHz
    D. 2.5 MHz
    E. 5.0 MHz

62. Suppose that we run a straight wire vertically through a horizontal piece of paper. We scatter iron filings on the paper, and then drive a steady, strong direct current through the wire. What will the iron filings do?
    A. Nothing; they'll remain arranged at random.
    B. They'll arrange themselves in radial lines centered on the point where the wire passes through the paper.
    C. They'll fly rapidly away from the wire and off the edges of the paper.
    D. They'll rush toward the point where the wire passes through the paper, and clump around that point in a tight little pile.
    E. None of the above

63. Suppose that we sweep all the iron filings off the paper in the experiment described in Question 62. We remove the wire and then place a bar magnet on the paper. Then we sprinkle some fresh iron filings onto the paper. What will the iron filings do?
    A. They'll fall onto the paper and remain arranged at random.
    B. They'll arrange themselves in straight radial lines, diverging from the middle of the bar magnet.

C. They'll fly rapidly away from the ends of the bar magnet and off the edges of the paper.
D. They'll arrange themselves in a pattern of concentric circles, centered at one end of the bar magnet.
E. None of the above

64. If we increase the frequency of an electromagnetic (EM) signal by a factor of 100, then its free-space wavelength
   A. becomes 10,000 times shorter.
   B. becomes 100 times shorter.
   C. does not change.
   D. becomes 100 times longer.
   E. becomes 10,000 times longer.

65. Suppose that we measure the resistance of a length of copper wire and find that it equals 0.025 ohms. What's the conductance of this wire sample in siemens?
   A. 0.025 S
   B. 2.5 S
   C. 4.0 S
   D. 25 S
   E. 40 S

66. Figure Exam-10 shows an AC transformer that has a tap in the secondary winding, with the turns counts specifically indicated. Suppose that we apply 120-V RMS utility AC, with no DC component, to the primary winding. What's the RMS voltage between terminals X and Y in the secondary?
   A. 21.3 V RMS
   B. 14.8 V RMS
   C. 12.0 V RMS
   D. 10.7 V RMS
   E. We need more information to say.

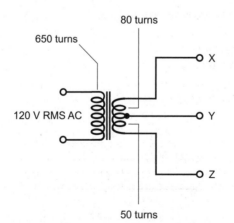

FIGURE EXAM-10 • Illustration for Final Exam Questions 66 through 70.

67. What's the RMS voltage between terminals Y and Z in the system of Fig. Exam-10?

   A. 9.2 V RMS
   B. 10.7 V RMS
   C. 11.1 V RMS
   D. 14.8 V RMS
   E. We need more information to say.

68. What's the RMS voltage between terminals X and Z in the system of Fig. Exam-10?

   A. 16.6 V RMS
   B. 18.0 V RMS
   C. 24.0 V RMS
   D. 27.0 V RMS
   E. We need more information to say.

69. What's the peak-to-peak voltage between terminals X and Z in the system of Fig. Exam-10?

   A. 67.9 V pk-pk
   B. 74.5 V pk-pk
   C. 80.0 V pk-pk
   D. 90.0 V pk-pk
   E. We need more information to say.

70. What's the ratio of the RMS voltage across the entire secondary winding to the RMS current through that same winding in the system of Fig. Exam-10?

   A. 25:1
   B. 5:1
   C. 1:5
   D. 1:25
   E. We need more information to say.

71. Figure Exam-11 illustrates an example of a

   A. full-wave rectifier.
   B. pi-section transformer.
   C. AC-to-DC converter.
   D. T-section power-supply filter.
   E. surge suppressor.

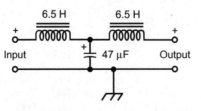

FIGURE EXAM-11 · Illustration for Final
Exam Questions 71 through 73.

72. **What should we expect to find at the input of the circuit shown in Fig. Exam-11, assuming that we use it in its intended role?**
    A. Pulsating DC
    B. Pure AC
    C. Pure DC
    D. AC with a DC component
    E. Any of the above

73. **What should we expect to find at the output of the circuit shown in Fig. Exam-11, assuming that we use it in its intended role?**
    A. Pulsating DC
    B. Pure AC
    C. Pure DC
    D. AC with a DC component
    E. Any of the above

74. **How well should we expect an open switch to conduct electricity?**
    A. Perfectly
    B. Very well
    C. Fairly well
    D. Poorly
    E. Not at all

75. **Figure Exam-12 on page 338 is a diagram of**
    A. an uninterruptible power supply.
    B. a voltage regulator circuit.
    C. a system for converting utility AC to higher-voltage AC.
    D. a photovoltaic power supply.
    E. a fuel-cell power supply.

76. **In the circuit of Fig. Exam-12, the box marked X represents**
    A. a full-wave bridge rectifier.
    B. a current limiter.
    C. a power inverter.
    D. an AC generator.
    E. a filter choke.

77. **Imagine that you want to build a long wire antenna for shortwave radio reception. The antenna must span 200 feet between two tall poles without any intervening supports. You plan to construct the antenna using stranded AWG No. 14 copper wire. The weight of the wire will put considerable mechanical stress on the span. You find a store that supplies 100-foot lengths of the wire you want, but no longer lengths. You'll have to splice the ends of two 100-foot lengths of wire together to make your antenna, and this splice will hang in the middle of the span. What type of splice will give you the best results in this situation?**
    A. An unsoldered twist splice
    B. A soldered twist splice

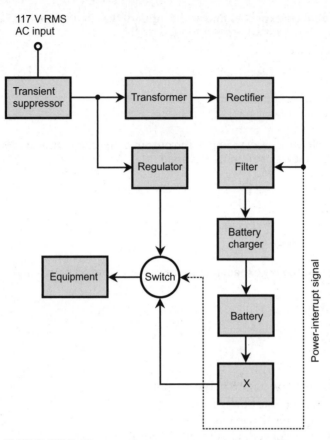

117 V RMS
AC input

FIGURE EXAM-12 • Illustration for Final Exam Questions 75 and 76.

C. An unsoldered Western Union splice
D. A soldered Western Union splice
E. It doesn't make any difference.

78. **If you vary the current through a component whose resistance remains constant, the voltage across that component**
   A. varies directly in proportion to the square of the current.
   B. varies directly in proportion to the current.
   C. does not change.
   D. varies in proportion to the reciprocal of the current.
   E. varies in proportion to the reciprocal of the square of the current.

79. **Parallel data cables sometimes suffer from crosstalk, a problem in which**
   A. signals in the various conductors interfere with each other.
   B. external EM fields upset the data transmission in the conductors.

C. the conductors radiate EM fields that interfere with external systems.

D. the individual conductors have different resistances, upsetting the balance of the entire system.

E. serial and parallel data streams get confused with each other.

80. **Imagine that you can magnetize a metal rod to a flux density of up to 1700 G, but no more than that, when you surround it with a coil carrying DC. When you remove the current from the coil, the residual flux density in the rod drops to 68 G. What's the retentivity of the rod?**

A. 4.0%

B. 5.0%

C. 16%

D. 25%

E. We need more information to calculate it.

81. **Figure Exam-13 illustrates a hypothetical deflection-versus-current curve for a compass galvanometer. What, if anything, is qualitatively wrong with this graph?**

A. It should be a straight line ramping down as you move from left to right.

B. It should be a straight line ramping upward as you move from left to right.

C. It should be a curve that decreases more and more gradually (rather than more and more rapidly, as shown here) as you move from left to right.

D. It should be a curve that increases more and more gradually as you move from left to right.

E. Nothing is wrong with this graph as shown.

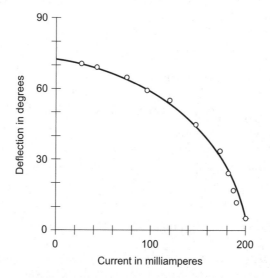

FIGURE EXAM-13 · Illustration for Final Exam Question 81.

82. Which of the following frequencies could represent ELF energy?

    A. 1.5 Hz
    B. 15 Hz
    C. 0.15 kHz
    D. 1.5 kHz
    E. All of the above

83. Imagine a long, thin, straight wire that produces a magnetic flux density of 8.00×10^{-7} T at any point 120 mm away from the wire. Suppose that the wire is suspended in free space, and no other objects are anywhere near it. What's the flux density at points 480 mm away from the wire?

    A. 1.00×10^{-7} T
    B. 2.00×10^{-7} T
    C. 4.00×10^{-7} T
    D. 8.00×10^{-7} T
    E. We need more information to say.

84. If we double the current through the wire in the scenario of Question 83 but we don't change the distance from the wire, what happens to the flux density at a particular point near the wire?

    A. It becomes half as great.
    B. It stays the same.
    C. It doubles.
    D. It quadruples.
    E. It increases by a factor of 8.

85. Suppose that we equip a mobile robot with a fluxgate magnetometer designed to allow the robot to navigate using the geomagnetic lines of flux for reference. We should *not* expect this device to perform well with the robot located

    A. near the geomagnetic equator.
    B. at any point not on the earth's surface.
    C. inside a steel-walled warehouse.
    D. over water.
    E. on board a fast-moving aircraft.

86. If we increase the free-space wavelength of an electromagnetic (EM) signal by a factor of 49, then its period

    A. becomes 2,401 times longer.
    B. becomes 49 times longer.
    C. becomes 7 times longer.
    D. does not change.
    E. None of the above

87. Maglev trains are typically propelled by

    A. linear motors.
    B. stepper motors.
    C. selsyns.
    D. synchros.
    E. relays.

88. **The term** *static electricity* **technically refers to a**
    A. potential difference without any flow of current.
    B. flow of current in the absence of any potential difference.
    C. lightning discharge that produces noise in a radio receiver.
    D. steady flow of direct current.
    E. steady flow of alternating current.

89. **Figure Exam-14 shows a logarithmic nomograph of the electromagnetic (EM) spectrum. The numbers at the top represent free-space wavelengths in meters. What can we correctly write in place of P?**
    A. Radio microwaves
    B. "Longwave" radio
    C. Infrared
    D. Visible light
    E. 60-Hz utility AC

90. **What can we correctly write in place of Q in Fig. Exam-14?**
    A. Visible light
    B. Infrared
    C. Alpha waves
    D. Radio microwaves
    E. Sound waves

91. **What can we correctly write in place of R in Fig. Exam-14?**
    A. Visible light
    B. Sound waves
    C. Ocean waves
    D. "Longwave" radio
    E. 60-Hz utility AC

92. **A superconductor can, at least in theory, expel magnetic flux altogether so that its permeability**
    A. increases to "infinity."
    B. decreases to 0.
    C. equals exactly 1.
    D. becomes negative.
    E. fluctuates at a high frequency.

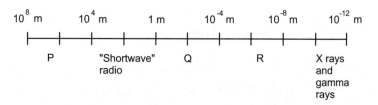

FIGURE EXAM-14 · Illustration for Final Exam Questions 89 through 91.

93. **We will observe a complete lack of magnetic inclination at any point on the earth's surface**
    A. on the geomagnetic equator.
    B. on a band around the earth connecting the geomagnetic poles.
    C. at or near the geographic poles.
    D. at or near the geomagnetic poles.
    E. whatsoever.

94. **Suppose that you reverse the polarities of all the cells in a four-cell flashlight. What will happen, assuming that electrical contact is maintained between the individual cells, and between the cells and the flashlight bulb?**
    A. The bulb will not glow at all.
    B. The bulb will glow, but less brightly than normal.
    C. The bulb will glow normally.
    D. The bulb will glow, but more brightly than normal.
    E. The bulb will burn out from excess current.

95. **If we increase the free-space wavelength of an electromagnetic (EM) signal by a factor of 49, then its frequency**
    A. becomes 2,401 times higher.
    B. becomes 49 times higher.
    C. becomes 7 times higher.
    D. does not change.
    E. None of the above

96. **If we connect the device shown in Exam-15 into the power-supply line for an electronic device so that terminals X and Z serve as the contacts of a switch, the device technically acts as**
    A. a normally open relay.
    B. an electromagnet.
    C. a commutator.
    D. a field coil.
    E. a chime.

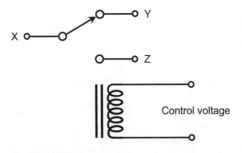

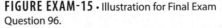

FIGURE EXAM-15 · Illustration for Final Exam Question 96.

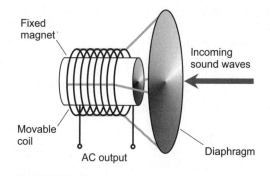

Fixed magnet

Incoming sound waves

Movable coil

Diaphragm

AC output

FIGURE EXAM-16 · Illustration for Final Exam Question 99.

97. **If we spin a nonferromagnetic, electrically conductive disk above an array of electromagnets fed with high-frequency AC, we can obtain**

A. negative feedback in the disk.

B. superconductivity in the disk.

C. levitation of the disk.

D. perpetual motion of the disk.

E. All of the above

98. **If we keep the voltage constant across a medium whose resistance changes, the current through that medium**

A. varies directly in proportion to the square of the resistance.

B. varies directly in proportion to the resistance.

C. does not change.

D. varies in proportion to the reciprocal of the resistance.

E. varies in proportion to the reciprocal of the square of the resistance.

99. **Figure Exam-16 is a functional diagram of a**

A. fluxgate magnetometer.

B. dynamic transducer.

C. selsyn.

D. chime.

E. synchro.

100. **If we bring a superconducting object that exhibits perfect diamagnetism near a permanent magnet, what will happen?**

A. The magnet will exert no force on the superconducting object.

B. One of the magnet's poles will attract the superconducting object, and the other magnetic pole will repel the object.

C. Both of the magnet's poles will attract the superconducting object.

D. Both of the magnet's poles will repel the superconducting object.

E. The object will lose its diamagnetic properties.

Answers to Quizzes, Tests, and Final Exam

Chapter 1	Chapter 3	Chapter 5	11. B
1. D	1. C	1. A	12. A
2. B	2. B	2. D	13. C
3. A	3. A	3. B	14. E
4. C	4. C	4. C	15. E
5. D	5. D	5. A	16. A
6. C	6. C	6. C	17. D
7. A	7. A	7. D	18. C
8. B	8. A	8. D	19. B
9. D	9. B	9. B	20. E
10. B	10. D	10. A	21. D
			22. A
Chapter 2	Chapter 4	Test: Part I	23. C
1. D	1. C	1. B	24. A
2. A	2. A	2. C	25. B
3. D	3. B	3. A	26. E
4. A	4. D	4. C	27. C
5. B	5. C	5. C	28. D
6. D	6. C	6. B	29. B
7. B	7. B	7. E	30. A
8. C	8. B	8. A	31. C
9. C	9. C	9. C	32. B
10. B	10. C	10. D	33. D

34. A
35. D
36. E
37. C
38. A
39. E
40. D
41. A
42. C
43. D
44. D
45. A
46. D
47. A
48. C
49. D
50. E

Chapter 6
1. C
2. C
3. C
4. B
5. A
6. D
7. D
8. B
9. C
10. A

Chapter 7
1. C
2. C
3. A
4. B
5. A
6. C
7. B

8. D
9. D
10. A

Chapter 8
1. A
2. C
3. A
4. B
5. C
6. B
7. D
8. A
9. D
10. D

Chapter 9
1. C
2. D
3. A
4. D
5. B
6. B
7. A
8. C
9. B
10. D

Test: Part II
1. E
2. E
3. C
4. C
5. A
6. D
7. E
8. B
9. A
10. C

11. D
12. B
13. C
14. B
15. B
16. E
17. C
18. D
19. A
20. E
21. B
22. C
23. A
24. A
25. C
26. D
27. A
28. D
29. E
30. E
31. E
32. D
33. C
34. B
35. B
36. C
37. A
38. B
39. C
40. E
41. E
42. C
43. A
44. D
45. C
46. A
47. B
48. E

49. D
50. A

Chapter 10
1. B
2. C
3. D
4. D
5. A
6. D
7. A
8. C
9. A
10. B

Chapter 11
1. C
2. A
3. B
4. B
5. B
6. B
7. D
8. A
9. B
10. D

Chapter 12
1. B
2. D
3. A
4. B
5. C
6. B
7. D
8. A
9. A
10. C

Test: Part III

1. E
2. C
3. D
4. C
5. B
6. C
7. A
8. B
9. A
10. D
11. E
12. E
13. C
14. E
15. B
16. B
17. C
18. B
19. A
20. C
21. A
22. E
23. B
24. A
25. A
26. B
27. C
28. D
29. E
30. B
31. D
32. C
33. E
34. B
35. D
36. C
37. E

38. A
39. E
40. C
41. A
42. C
43. B
44. D
45. A
46. E
47. B
48. E
49. C
50. B

Final Exam

1. B
2. E
3. B
4. E
5. B
6. D
7. A
8. A
9. E
10. C
11. B
12. E
13. A
14. B
15. D
16. D
17. B
18. C
19. A
20. C
21. E
22. E
23. B
24. C

25. E
26. A
27. B
28. C
29. A
30. C
31. D
32. A
33. C
34. E
35. A
36. D
37. B
38. A
39. E
40. D
41. D
42. A
43. B
44. D
45. E
46. D
47. C
48. D
49. C
50. A
51. B
52. D
53. C
54. D
55. D
56. B
57. C
58. A
59. D
60. B
61. E
62. E

63. E
64. B
65. E
66. B
67. A
68. C
69. A
70. E
71. D
72. A
73. C
74. E
75. A
76. C
77. D
78. B
79. A
80. A
81. D
82. E
83. B
84. C
85. C
86. B
87. A
88. A
89. E
90. D
91. A
92. B
93. A
94. C
95. E
96. A
97. C
98. D
99. B
100. D

Schematic Symbols

ammeter

amplifier, general

amplifier, inverting

amplifier, operational

AND gate

antenna, balanced

antenna, general

antenna, loop

antenna, loop, multiturn

battery, electrochemical

capacitor, feedthrough

capacitor, fixed

capacitor, variable

capacitor, variable,
   split-rotor

capacitor, variable,
   split-stator

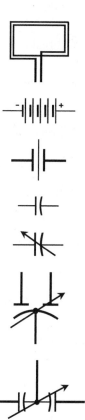

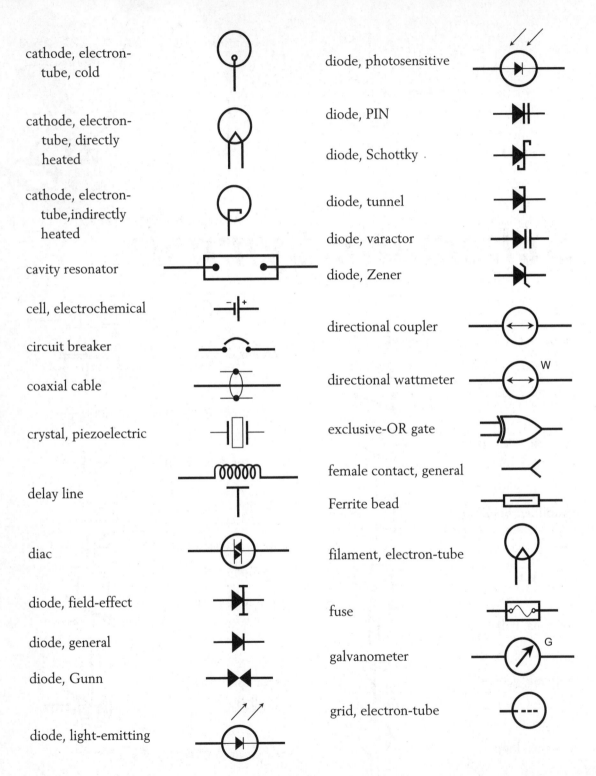

cathode, electron-tube, cold

cathode, electron-tube, directly heated

cathode, electron-tube, indirectly heated

cavity resonator

cell, electrochemical

circuit breaker

coaxial cable

crystal, piezoelectric

delay line

diac

diode, field-effect

diode, general

diode, Gunn

diode, light-emitting

diode, photosensitive

diode, PIN

diode, Schottky

diode, tunnel

diode, varactor

diode, Zener

directional coupler

directional wattmeter

exclusive-OR gate

female contact, general

Ferrite bead

filament, electron-tube

fuse

galvanometer

grid, electron-tube

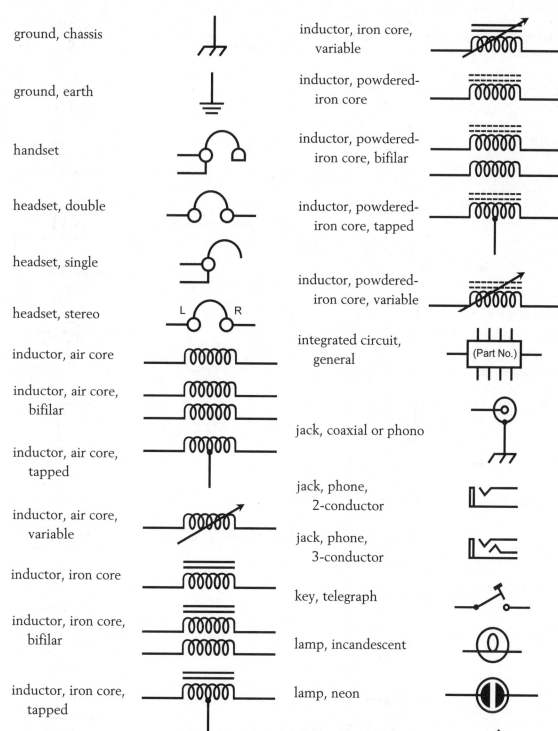

ground, chassis	inductor, iron core, variable
ground, earth	inductor, powdered-iron core
handset	inductor, powdered-iron core, bifilar
headset, double	inductor, powdered-iron core, tapped
headset, single	inductor, powdered-iron core, variable
headset, stereo	integrated circuit, general
inductor, air core	jack, coaxial or phono
inductor, air core, bifilar	jack, phone, 2-conductor
inductor, air core, tapped	jack, phone, 3-conductor
inductor, air core, variable	key, telegraph
inductor, iron core	lamp, incandescent
inductor, iron core, bifilar	lamp, neon
inductor, iron core, tapped	male contact, general

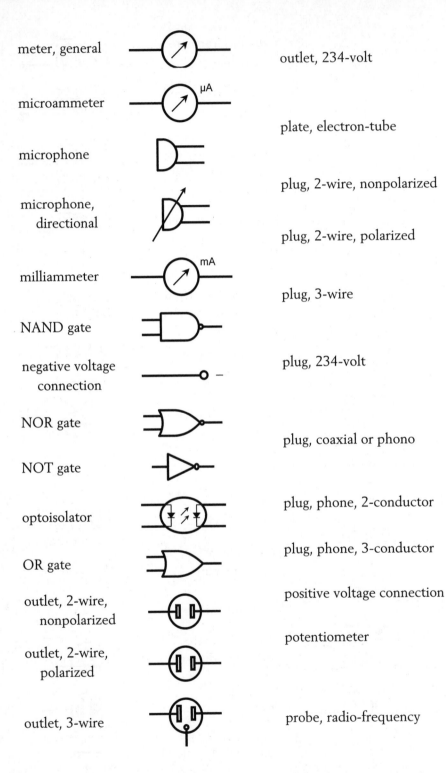

meter, general

microammeter

microphone

microphone, directional

milliammeter

NAND gate

negative voltage connection

NOR gate

NOT gate

optoisolator

OR gate

outlet, 2-wire, nonpolarized

outlet, 2-wire, polarized

outlet, 3-wire

outlet, 234-volt

plate, electron-tube

plug, 2-wire, nonpolarized

plug, 2-wire, polarized

plug, 3-wire

plug, 234-volt

plug, coaxial or phono

plug, phone, 2-conductor

plug, phone, 3-conductor

positive voltage connection

potentiometer

probe, radio-frequency

or

rectifier, gas-filled

rectifier, high-vacuum

rectifier, semiconductor

rectifier, silicon-controlled

relay, double-pole,
double-throw

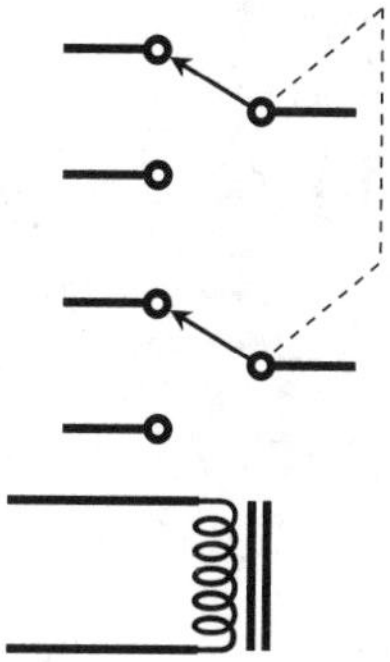

relay, double-pole,
single-throw

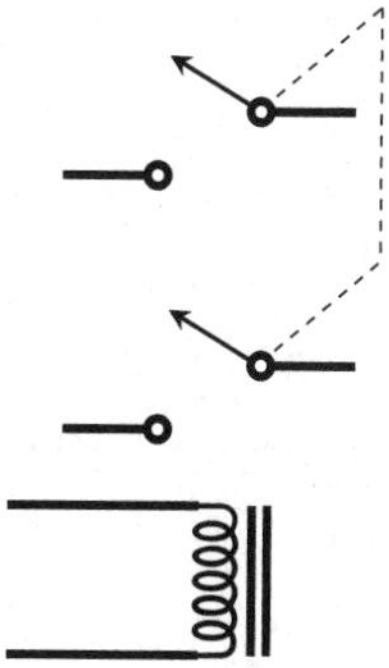

relay, single-pole,
double-throw

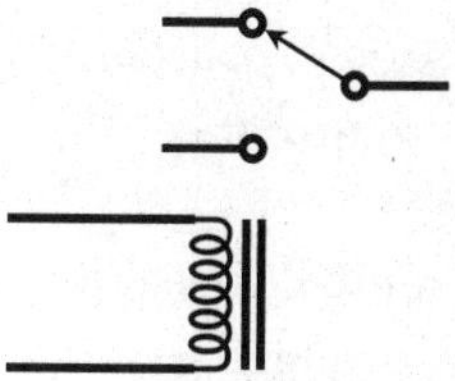

relay, single-pole,
single-throw

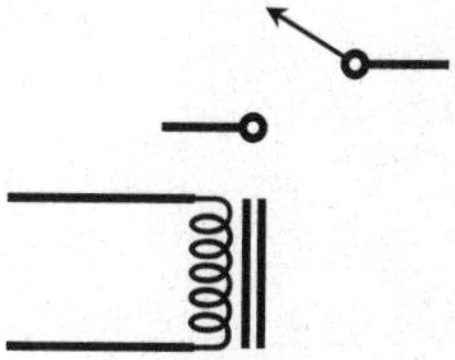

resistor, fixed

resistor, preset

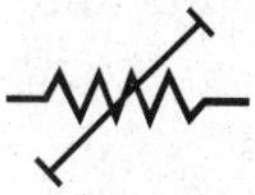

resistor, tapped

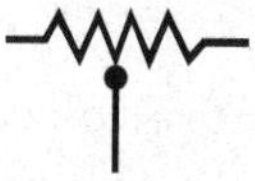

resonator

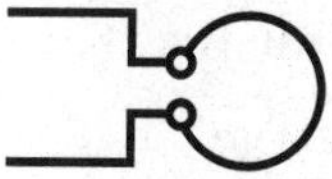

rheostat

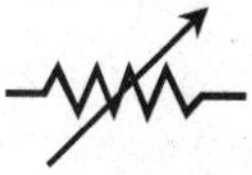

saturable reactor

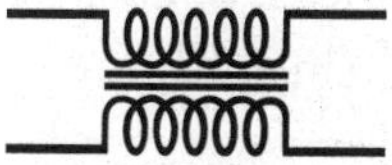

signal generator

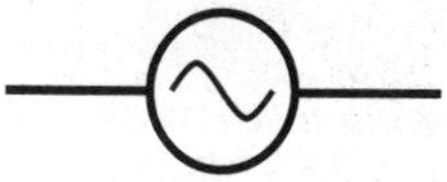

solar battery

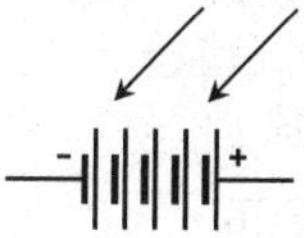

solar cell

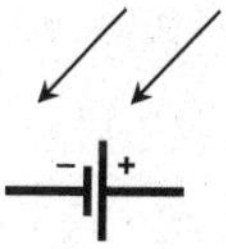

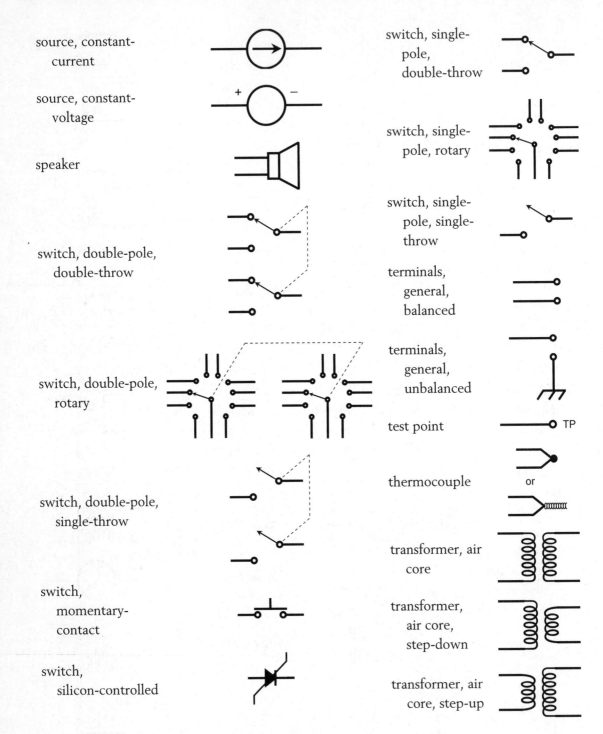

source, constant-current

source, constant-voltage

speaker

switch, double-pole, double-throw

switch, double-pole, rotary

switch, double-pole, single-throw

switch, momentary-contact

switch, silicon-controlled

switch, single-pole, double-throw

switch, single-pole, rotary

switch, single-pole, single-throw

terminals, general, balanced

terminals, general, unbalanced

test point

thermocouple   or

transformer, air core

transformer, air core, step-down

transformer, air core, step-up

transformer, air core, tapped primary

transformer, powdered-iron core, tapped secondary

transformer, air core, tapped secondary

transistor, bipolar, NPN

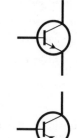

transformer, iron core

transistor, bipolar, PNP

transformer, iron core, step-down

transistor, field-effect, N-channel

transformer, iron core, step-up

transistor, field-effect, P-channel

transformer, iron core, tapped primary

transistor, MOS field-effect, N-channel

transformer, iron core, tapped secondary

transistor, MOS field-effect, P-channel

transformer, powdered-iron core

transistor, photosensitive, NPN

transformer, powdered-iron core, step-down

transistor, photosensitive, PNP

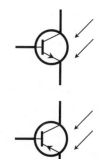

transformer, powdered-iron core, step-up

transformer, powdered-iron core, tapped primary

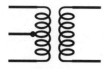

transistor, photosensitive, field-effect, N-channel		tube, photosensitive	
transistor, photosensitive, field-effect, P-channel		tube, tetrode	
transistor, unijunction		tube, triode	
triac		unspecified unit or component	
tube, diode		voltmeter	
		wattmeter	
tube, heptode		waveguide, circular	
		waveguide, flexible	
tube, hexode		waveguide, rectangular	
		waveguide, twisted	
tube, pentode			

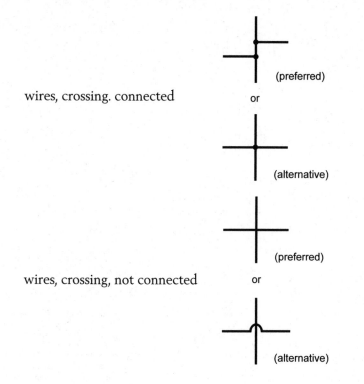

wires, crossing. connected   or

(preferred)

(alternative)

(preferred)

wires, crossing, not connected   or

(alternative)

Suggested Additional Reading

Gerrish, Howard, *Electricity and Electronics*. Tinley Park, IL: Goodheart-Wilcox Co., 2008.

Gibilisco, Stan, *Algebra Know-It-All*. New York, NY: McGraw-Hill, 2008.

Gibilisco, Stan, *Electricity Experiments You Can Do at Home*. New York, NY: McGraw-Hill, 2010.

Gibilisco, Stan, *Electronics Demystified*, 2nd ed. New York, NY: McGraw-Hill, 2011.

Gibilisco, Stan, *Physics Demystified*, 2nd ed. New York, NY: McGraw-Hill, 2011.

Gibilisco, Stan, *Pre-Calculus Know-It-All*. New York, NY: McGraw-Hill, 2010.

Gibilisco, Stan, *Teach Yourself Electricity and Electronics*, 5th ed. New York, NY: McGraw-Hill, 2011.

Gibilisco, Stan, *Technical Math Demystified*. New York, NY: McGraw-Hill, 2006.

Gibilisco, Stan, and Crowhurst, Norman, *Mastering Technical Mathematics*, 3rd ed. New York, NY: McGraw-Hill, 2007.

Gottlieb, I. M., *Electric Motors and Control Techniques*, 2nd ed. New York, NY: TAB/McGraw-Hill, 1994.

Gussow, Milton, *Schaum's Outline of Basic Electricity*, 2nd ed. New York, NY: McGraw-Hill, 2009.

Morrison, Ralph, *Electricity: A Self-Teaching Guide*, 3rd ed. Hoboken, NJ: John Wiley & Sons, Inc., 2003.

Slone, G. Randy, *TAB Electronics Guide to Understanding Electricity and Electronics*, 2nd ed. New York, NY: McGraw-Hill, 2000.

Index